Elaheh Mousavi

Urban Virtual Reality Models for Engineering Applications

Elaheh Mozaffari

Urban Virtual Reality Models for Engineering Applications

Creating and Testing Models

VDM Verlag Dr. Müller

Bibliographic information by the German National Library: The German National Library lists this publication at the German National Bibliography; detailed bibliographic information is available on the Internet at http://dnb.d-nb.de.

Contact: info@vdm-verlag.de
Cover image: www.purestockx.com
Publisher: VDM Verlag Dr. Müller e. K., Dudweiler Landstr. 125 a, 66123 Saarbrücken, Germany
Produced by: Lightning Source Inc., La Vergne, Tennessee/USA
        Lightning Source UK Ltd., Milton Keynes, UK

Bibliografische Information der Deutschen Nationalbibliothek: Die Deutsche Nationalbibliothek verzeichnet diese Publikation in der Deutschen Nationalbibliografie; detaillierte bibliografische Daten sind im Internet über http://dnb.d-nb.de abrufbar.

Kontakt: info@vdm-verlag.de
Coverbild: www.purestockx.com
Verlag: VDM Verlag Dr. Müller e. K., Dudweiler Landstr. 125 a, 66123 Saarbrücken, Deutschland
Herstellung: Lightning Source Inc., La Vergne, Tennessee/USA
        Lightning Source UK Ltd., Milton Keynes, UK

ISBN: 978-3-8364-2349-6

ABSTRACT

Creation and Testing Urban Virtual Reality Models for Engineering
Eleheh Mozaffari

Virtual Reality interaction methods can provide better understanding of 3D virtual models of urban environments. These methods can be used to accommodate different civil engineering applications, such as facilities management and construction progress monitoring. However, the users of these applications may have severe problems in exploring and navigating the large virtual environments to accomplish specific tasks. Properly designed navigation methods are critical for using these applications efficiently. In this reserch, first a literature survey is conducted about existing navigation and visualization methods in order to identify the methods that are most suitable and practical for engineering applications. Based on this survey, a taxonomy of navigation methods and support tools for engineering applications in urban virtual environments is developed. In addition, a framework for virtual reality applications in civil engineering is proposed. The framework includes a practical method for creating virtual urban models and several interaction and navigation methods. Furthermore, a new method is proposed for usability testing of the navigation supports in these models. The proposed approach is demonstrated through three case studies for desktop, indoor and outdoor applications. The results of our usability study showed that using navigation supports allow users to navigate more efficiently in the virtual environment.

ACKNOWLEDGMENTS

I would like to thank Professor Amin Hammad for being my mentor. His profound knowledge and thoughtful instructions have always shed some light on my way to pursue this work.

I also wish to express my special gratitude to my husband Professor Javad Dargahi, and my parents for being enduring, encouraging and caring to my aspirations. I thank them for inspiring me to accomplish this work. To them, I dedicate this book.

TABLE OF CONTENTS

LIST OF FIGURES

LIST OF TABLES

LIST OF ABBREVATIONS

Abbreviation	Description
2D	Two-dimensional
3D	Three-dimensional
ANOVA	Analysis of Variances
AR	Augmented Reality
BMS	Bridge Management System
CAAD	Computer- Aided Architectural Design
CAVE	Cave Automatic Virtual Environment
DB	Database
DEM	Digital Elevation Model
FM	Facilities Management
FMS	Facilities Management System
GIS	Geographic Information System
GPS	Global Positioning System
GUI	Graphical User Interface
HMD	Head-Mounted Display
I/O	Input/Output
JDBC	Java Database Connectivity
LBC	Location-Based Computing
LBC-Infra	Location-Based Computing for Infrastructure field tasks
LoD	Levels of Detail
MR	Mixed Reality
TIN	Triangulated Irregular Network
UVE	Urban Virtual Environment
VE	Virtual Environment
VR	Virtual Reality
VRML	Virtual Reality Modeling Language
WoW	Window on World Systems

CHAPTER 1   INTRODUCTION

1.1 GENERAL BACKGROUND

With recent developments in computer graphics performance and new software technologies, we experience a switch in architectural visualizations from still images and pre-rendered animations to interactive 3D models. Virtual Reality (VR) is one of the techniques for representing 3D spatial data. It enables complex details of the real world to be visualized by utilizing the human ability to navigate through familiar environments and fully interact with spatial information of various types.

3D city models have been useful for many urban applications such as planning and construction. However, in order to realize 3D urban models, enormous amount of time and money has to be invested to design the models and to acquire the necessary data. Large-scale urban virtual environments (UVEs) can be created using commercial software. However, these models focus only on the exterior shapes of buildings and do not include all the details necessary in engineering applications (e.g., design and scheduling information). Therefore, a practical and economic data integration method for creating the 3D model of an urban environment is needed.

Another issue related to UVEs is that the users of these applications may have severe problems in exploring and navigating the large virtual environments to accomplish specific tasks. For accomplishing their tasks in these environments, they need to be able to navigate either to accomplish specific tasks or to become more familiar with the environments. Properly designed user interfaces for navigation can make that experience successful and enjoyable. There are many research works on the general principles of 3D navigation and wayfinding in Virtual Environments (VEs). However, little work has been done about providing navigation support in engineering applications involving large scale urban environments.

1.2 OBJECTIVES

In this book a framework for VR applications in civil engineering is investigated. The framework includes a practical method for creating UVEs and several interaction and navigation methods suitable for VR applications in civil engineering. The objectives of this book are as follows:

(1) To investigate a framework for VR applications in civil engineering;

(2) To investigate a practical method for creating UVEs by integrating multiple data sources;

(3) To investigate the interaction and navigation methods suitable for the above UVEs applications;

(4) To investigate an approach for usability testing of interaction and navigation methods in the above UVEs.

1.3 ORGANIZATION OF THE BOOK

This study will be presented as follows:

Chapter 2 Literature Review: This chapter presents the current situation of VR and its applications in urban environments. Several approaches for creating virtual cities are introduced and usability testing of interaction and navigation in VEs are reviewed. In addition, mobile Location-Based Computing (LBC) technologies and its application in civil engineering are introduced.

Chapter 3 Framework for virtual reality applications in civil engineering: In this chapter, UVEs are discussed within a larger framework. The generic structure of this framework embodies the general functionalities of VR applications in civil engineering and can be used to suit the requirement of mobile infrastructure management systems. A data integration method is proposed for creating the UVEs by synthesizing information from different data sources. In addition, three tracking methods used in this framework are introduced.

Chapter 4 Interaction in urban virtual environments and usability metrics: This chapter introduces the interaction and navigation components of the proposed framework and a taxonomy for navigation methods and support tools in UVEs for civil engineering applications. In addition, a GIS-based evaluation technique for usability testing of navigation is discussed.

Chapter 5 Implementation and Case Studies: In this chapter, several case studies are used to demonstrate the prototype system using the proposed approaches.

Chapter 6 Summary, Conclusions, and Future work: This chapter summarizes the present research work, highlights its contributions, and suggests recommendations for future research.

CHAPTER 2 LITERATURE REVIEW

2.1 VIRTUAL REALITY AND VIRTUL ENVIRONMENS

VR is a technology that creates a virtual three-dimensional (3D) model in a computer to visually reproduce the shape, texture and movement of objects (Burdea and Coiffet, 2003). The environment where this model exists is called as the VE. The VE may be a model of a real-world object, such as a house; it might be an abstract world that does not exist in a real sense but is understood by humans, such as a chemical molecule or a representation of a set of data; or it might be in a completely imaginary science fiction world. A key feature of VEs is that the user believes that he/she is actually in this different world. A second key feature is that if the human moves his head, arms or legs, the shift of visual cues must be those he/she would expect in a real world. In other words, besides immersion, VEs should support navigation and interaction. Different kinds of VE technology support different modes of interaction (Bowman and Billinghurst, 2002).

2.1.1 Types of VR systems

Isdale (2003) categorized VR systems by the way with which they interface to the user. The following are some of the common modes used in VR systems:

(1) Window on World Systems (WoW) or Desktop VR

Desktop VR systems use a conventional computer monitor to display the virtual world. The user views the system from the usual close but remote position and interacts through standard or special-purpose input or control devices such as keyboards, mouse controls, trackballs, and joysticks (Figure 2.1).

Figure 2.1 Desktop VR system (Panoram Technologies, 2006)

(2) Video Mapping

A variation of the WoW approach merges a video input of the user's silhouette with a 2D computer graphic. Video Mapping VR uses cameras to project an image of the user into the computer program, thus creating a 2D computer character. The user watches a monitor that shows his body's interaction with the world.

(3) Immersive Systems

The ultimate VR systems completely immerse the user's personal viewpoint inside the virtual world. These "immersive" VR systems are often equipped with a Head-Mounted Display (HMD) (Figure 2.2 (a)). An HMD houses two small display screens and an optical system that channels the images from the screens to the eyes, while a motion tracker continuously lets an image-generating computer adjust the scene to the user's current view (Ausburn and Ausburn, 2004). As a result, the viewer can look around and walk through the surrounding VE. A variation of the immersive system is the CAVE

(Cave Automatic Virtual Environment) (Figure 2.2 (b)). The CAVE provides the illusion of immersion by projecting stereo images on the walls and floor of a room-sized cube. Several persons wearing HMDs or lightweight stereo glasses can enter and walk freely inside the CAVE. A head tracking system continuously adjusts the stereo projection to the current position of the leading viewer (Virtual Reality Laboratory, 2006).

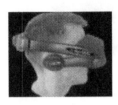

(a) HMD (stereo3d, 2006)   (b) CAVE immersive systems
                                         (UM3D, 2006)

Figure 2.2 Immersive VR systems

(4) Telepresence

Using telepresence, *u*sers can influence and operate in a world that is real but in a different location. The users can observe the current situation with remote cameras and achieve actions via robotic and electronic arms. Telepresence is used for remote surgical operations (Garner et al., 1997) and for the exploration/manipulation of hazardous environments such as the space (Goza et al., 2004), underwater and radioactive environments.

(5) Mixed Reality (MR)

Mixed Reality (MR) was introduced by Milgram and Kishino (1994), where different combinations of the virtual and real components are possible along a virtuality continuum (Figure 2.3). The real world and a totally VE are at the two ends of this continuum with the middle region called MR. Augmented Reality (AR) lies near the real world end of the line with the perception of the real world augmented by computer generated data. This technology combines the viewing of the real-world or video-based environments with superimposed 3D virtual objects that can be manipulated by the viewer. Thus, AR supplements rather than replaces the user's real world (Virtual Reality Laboratory website, 2006). The most recent advancement in AR is a wearable system in which users wear a backpack with a portable computer, see-through HMD, and headphones with motion trackers to place and manipulate virtual objects as they move within their real world (Halden Virtual Reality Center, 2006). Augmented Virtuality (AV) is a term created by Milgram and Kishino (1994) to identify systems which are mostly synthetic with some real world imagery added, such as texture mapping video, onto virtual objects.

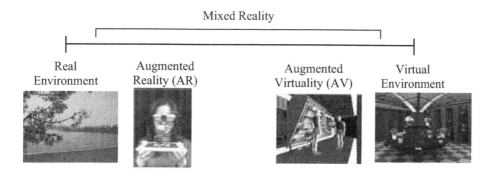

Figure 2.3 MR continuum (Isdale, 2003)

2.1.2 Applications of virtual reality

VR arguably provides the most natural means of communicating with a computer because it allows users to use their inherent 3D spatial skills, such as walking, gesturing, looking, grabbing etc. There is a vast range of potential applications of VR including the following applications:

(1) Visualization and representation tool

The interactive nature of VEs makes them a natural extension to 3D graphics that enable engineers, architects, and designers to visualize real-life structures before actually building them. Another advantage of VEs over other computer-based design tools is that they enable the user to visualize models that are difficult to understand in other ways. It can therefore be a valuable tool for a chemist exploring ways that molecules can be combined (Cruz-Neira et al., 1996), or a medical researcher trying to forecast the effects of complex combinations of healing treatments. Furthermore, a VE can be made to respond to actual or hypothetical physical laws. Gravity, for example, can be suspended.

(2) Distance training and communication

In distance training and communication, separate computers are joined together through the Internet or other computer networks. The remote participants share the representation of the same virtual world and interact with one another. VEs are currently being used as conferencing tools on the Internet and they do give users a sense of "being there". Of 53 participants who had attended one such Internet-based conference (scheduled scientific discussions over a four-week period), 74 % reported feeling a sense of actually being in the same room with others (Towell and Towell, 1995).

(3) Hands-on training

VR is currently used to train operators of various kinds of equipment, where initial training in a VE can avoid the expense, danger, and problems of monitoring and control associated with training in real-life situations. For example, VR can be used to train individuals to perform tasks in dangerous environments, such as in radioactive emergencies or icy road conditions. In addition to the assurance of safety, the use of a training VE gives the trainer total control over many aspects of the trainee's performance. The best-known example of VR hands-on training is flight simulation, but VR training is increasingly being used in the initial training of divers, surgeons (Hoffman and Vu, 1997) and shipboard fire-fighters (Tate et al., 1998).

(4) Orientation and navigation

VR is well suited to helping users learn to navigate in unfamiliar or complex surroundings. A number of studies have shown that navigation in VR models of complex buildings can help understanding the configuration of actual buildings (Goldberg, 1994). Other studies have found that military personnel using self-guided virtual terrain environments can learn successfully the actual physical terrain that had been simulated (Johnson, 1994). Researchers in Japan are developing a system to train guards of power plants in navigating their rounds. In this application, the VR training takes place before workers are assigned to the site, when the workers are still unfamiliar with the look and layout of an actual plant (Umeki and Doi, 1997).

2.2 VIRTUAL CITIES

The needs for 3D city models are growing and expanding rapidly in various fields including urban planning and design, architecture and environmental visualization. The efficient generation of 3D city models is improving the practice of urban environmental planning and design. For example, planning authorities will be able to illustrate explicit photo-textured information of what the city environment will look like after a proposed change. Photo-textured and 3D models enable easy understanding of cities. The computational power of this technology to transform and instantly compare alternative representations provides decision-makers with unprecedented flexibility. When and if visualization tools and good data are widely available, one will be able to propose and show design changes to a city.

There are various terms used for 3D city models, such as *Virtual City*, *Cybercity*, and *Digital City* (Dodge et al., 1998). 3D city models are basically computerized or digital models of a city including the graphical representation of buildings and other objects in 2.5D or 3D aspects. The term 2.5D is used for describing models where there is only one unique Z-value (elevation value) defined for each pair of XY-coordinates (Sinning-Meister et al., 1996). Although CAD models or solid models can represent 3D, the highest form of visualization of 3D city models is using VR because of the easiness of navigating through the model.

2.2.1 Applications for 3D Urban Models

Several 3D city models were developed for a wide range of applications. Batty et al. (2000) classified these applications into 12 different categories. Shiode (2001), regrouped this classification into four different general categories of use: (1) planning and design; (2) infrastructure and facility services; (3) commercial sector and marketing; and (4) promotion and learning of information on cities.

(1) Planning and design

Planning and detailed design reviews as well as problems of site location, community planning and public participation all require and are informed by 3D visualization. Visual representation of environmental impact is also widely supported by 3D models. This concerns various kinds of hazard to be visualized and planned for, and ways of visualizing the impact of disasters as well as local pollutants.

(2) Infrastructures and facility services

Urban infrastructure such as water, sewers, and electricity provision, as well as road and rail networks, all require detailed 2D and 3D data for their improvement and maintenance. The analysis of line-of-sight for mobile and fixed communications is also crucial in environments dominated by high-rise buildings to secure a clear reception of signals.

(3) Commercial sector and marketing

2D and 3D models are effective for visualizing spatial distribution of the clients and market demands for specific economic activities as well as the availability of space for development. They also enable the computation of detailed data concerning floor-space and land availability as well as land values and costs of development. Finally, virtual city models in 2D and 3D provide portals to virtual commerce through semi-realistic entries to remote trading and other commercial domains.

(4) Promotion and learning of information on cities

3D visualization offers entries to urban information hubs where users at different levels of education can learn about the city and access other learning resources through the metaphor of the city. In particular, it provides methods for displaying the tourist attractions of cities as well as ways in which tourists and other newcomers can learn about the geography of the city.

2.2.2 Examples of 3D city models

One of the earliest examples of 3D city model is the Bath Model (Bourdakis and Day, 1997) shown in Figure 2.4. This model was developed by the team of researcher at the Centre for Advanced Studies in Architecture (CASA) at the University of Bath.

A review of 3D city models (Batty et al., 2001) stated that there are over 60 large-scale 3D city model projects worldwide each of which modeling a part of an existing city (Figure 2.5). The two best developed sets of applications that they found are in Tokyo (Figure 2.6) and New York (Figure 2.7). Most of the 3D city models were being developed for a very wide range of applications.

Figure 2.4 The Bath model (Bourdakis and Day, 1997)

(a) Virtual Los Angeles   (b) Virtual New York
(Urban simulation team at UCLA, 2006)   (Planet 9 Studio, 2006)

(c) Virtual Tokyo (Planet 9 Studio, 2006)   (d) Virtual Bologna, Italy (NUME, 2006)

Figure 2.5 Examples of virtual cities

(a) The Asian air survey   (b) Fire spread model
model   (Zenrin, 2006)

(c) Digital panoramic visualization (Webscape, 2006)

Figure 2.6 Virtual City of Tokyo with three different applications (Batty et al., 2000)

(a) Floorspace in downtown
Manhattan (UDS, 2006)

(b) Floorspace (ESC, 2006)

(c) Draping orthophotos
on 3D block models (ASI, 2006)

(d) Utility lines beneath the
floorspace blocks (ASI, 2006)

Figure 2.7 Virtual City of New York (Batty et al., 2000)

2.2.3 Virtual Reality and Geographic Information Systems

Advancement of information technology provides a wide array of tools that support both geographical analysis and urban modeling. As a result of these advancements, mass scale models in 3D have been appearing on the World-Wide Web (Google Earth, 2006). 3D visualization techniques have made complex geometry quite simple to model based on new methods of imaging and remote sensing, thus improving the quality of rendering. In the recent past, models were based on Computer-Aided Architectural Design (CAAD) where accurate geometry was considered critical. Most of these failed to provide spatial data analysis until Geographic Information Systems (GISs) technology was developed

14

and made available. Developments in geomatic engineering and GISs meet demands for querying spatial data structures and visualizing the results.

By adding the dimension of height to building footprints, it is possible to create simple 3D city models, which can be embellished with attribute information in a way similar to GIS operations in 2D thematic data (Faust, 1995; Levy, 1995). The traditional mapping and database functionalities of GISs are being augmented with an array of rich multimedia data (Shiffer, 1995) which have certainly added value to the way GISs is used in planning, urban design and as a decision-making tool in general.

Applied researches (Smith, 1997; Martin and Higges, 1997) and commercial activities (Gross and Kennelly, 2005) have made advances in the field of VR-GIS by making it easier to generate 3D urban models from 2D spatial and thematic data. This will allow the planners to easily build large 3D city models on their desktops.

2.2.4 Modeling methods of 3D city models

There are a number of factors in 3D urban models, such as clients, applications, the output expected, budget, time period, and the amount of area to be covered. Shiode (2001) proposes three factors which in turn will dictate the technique or method of construction used for the large scale 3D city models: (1) the degree of reality; (2) types of data input; and (3) the degree of functionality and the ability to conduct various analyses. Figure 2.8 shows a summary of models in terms of the difference in geometrical details (Shiode, 2001): (1) 2D digital maps and digital ortho-photographs: Conventional 2D GIS maps support a range of applications but are incapable of giving an intuitively comprehensive 3D representation; (2) Image-based rendering: Panoramic image-based modeling (Ishiguro, 2003) is an inexpensive solution for psuedo-3D visualization, although the number of shots taken will limit its viewpoints, and it would certainly not incorporate spatial analysis functions; (3) Prismatic building block models: Block extrusion is a fusion of 2D building footprints with airborne survey data and other height resources. GIS technology allows overlying 2D maps on airborne data and determining the spatial characteristics of the image within each building footprint. Prismatic building block models lack the architectural details and convey no compelling sense of the environment, but are sufficient for analyzing view sheds and the shortest path; (4) Block modeling with image-based texture mapping: These are similar to the prismatic building block models but with image-based facades. The building textures are most commonly generated from either oblique aerial or terrestrial images, which in most cases successfully compensate for the simplification of the outline of building geometry and roof shapes (Batty et al., 2000); (5) Models with architectural details and roof shapes: Modern digital photogrammetric systems enable an efficient recovery of 3D surface details. Automated search techniques are used for identifying the corresponding locations (points, edges and regions) in multiple, overlapping images to generate a number of possible geometries which can be tested against templates but still require significant manual intervention for architecturally rich contents (Groneman, 2004); and (6) Full volumetric CAD models: As-built CAD models of individual buildings are frequently undertaken by a combination of measured building survey and terrestrial photogrammetry. The complexity of such models range from digital orthophotographs (in which images are rectified and combined to remove perspective effects) to the full architectural details, but the cost would be prohibitively expensive for full city coverage.

In terms of application and market demands, perhaps the most crucial feature is the functionality. Photo-realistic CAD-type models are often less functional whereas GIS-based models are generally supported by substantial attribute data and are integral to some analysis. While the amount of analytical features does not necessarily determine the usefulness of a model in its specific context, the potential for extensive and alternative use is directly reflected where GIS functionality is available.

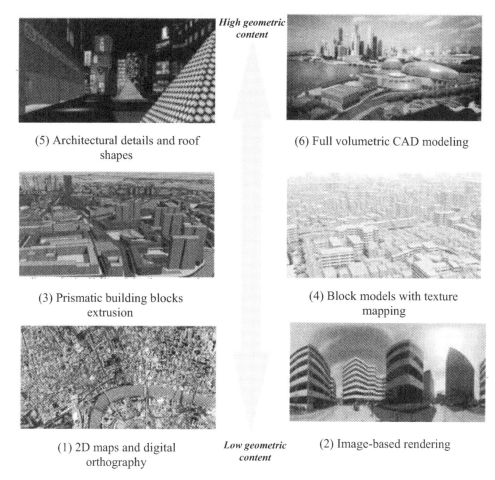

High geometric content

(5) Architectural details and roof shapes

(6) Full volumetric CAD modeling

(3) Prismatic building blocks extrusion

(4) Block models with texture mapping

(1) 2D maps and digital orthography

Low geometric content

(2) Image-based rendering

Figure 2.8 3D urban models using different modeling methods (Shiode, 2001)

17

2.3 LOCATION-BASED COMPUTING FOR ENGINEERING FIELD TASKS

Location Based Computing (LBC) is an emerging discipline focusing on integrating geoinformatics, telecommunications, and mobile computing technologies. LBC utilizes geoinformatics technologies, such as GISs and the Global Positioning System (GPS) in a distributed real-time mobile computing environment. LBC is paving the way for a large number of location-based services and is expected to become pervasive technology which people will use in daily activities, such as mobile commerce, as well as in critical systems, such as emergency response systems. In LBC, elements and events involved in a specific task are registered according to their locations in a spatial database, and the activities supported by the mobile and wearable computers are aware of these locations using suitable positioning devices. For example, an inspection system based on LBC would allow the bridge inspector to accurately locate the defects on a predefined 3D model of the bridge in real time without the need for any post-processing of the data. Hammad et al. (2004a) discussed the concept of a mobile data collection system for engineering field tasks based on LBC called *LBC for Infrastructure field tasks* or LBC-Infra. The concept of LBC-Infra is to integrate spatial databases, mobile computing, tracking technologies and wireless communications in a computer system that allows infrastructure field workers using mobile and wearable computers to interact with georeferenced spatial models of the infrastructure and to automatically retrieve the necessary information in real time based on their location, orientation, and specific task context. Using LBC-Infra, field workers will be able to access and update information related to their tasks in the field with minimum efforts spent on the interaction with the system, which results in increasing their efficiency and reducing the cost of infrastructure inspection. In addition, the wirelessly distributed nature of LBC will allow field workers to share the collected information and communicate with each other and with personnel at a remote site (office). This feature is of a great value especially in emergency cases.

2.4 MOBILE AND LOCATION- BASED COMPUTING ISSUES

As we discussed, LBC is based on tracking the location of the user and providing him/her with information based on this location in a distributed mobile computing environment (Satyanarayanan, 2001; Davies et al., 2001). For example, data collection for a Facilities Management Systems (FMS) or a Bridge Management System (BMS) can be accomplished using a laptop equipped with a tracking device so that dispatchers of work orders can view where workers are located in real time and determine the closest to handle an emergency repair. In addition, the virtual model acts as a guide for the inspectors to locate the infrastructure and every part of it. Using this approach enables users to interact with the environment, which significantly enhances their understanding of the environment.

An important component of LBC is the mobile computing platforms. These platforms, such as Tablet PCs are being developed with integrated barcode technology and wireless communications for applications such as real-time inventory and work order information. Advanced data collection terminals will allow individuals to download, view, and update schedules, update work orders, issue and receive inventory items, track product movements, and communicate with personnel in operations, maintenance, or corporate.

Figure 2.9 shows the concept of LBC-Infra (Hammad et al., 2004a). In this figure, a bridge inspector, equipped with a wearable computer, is inspecting a large highway bridge searching for damages, such as cracks. The inspector is equipped with a mobile or wearable computer that has a wireless communications card and is connected to tracking devices. Based on the location and orientation of the inspector and the task to be achieved, the system may display information about the parts of interest within his or her focus or navigation arrows to the locations where cracks are most likely to be found. The inspector compares the changes in conditions by wirelessly accessing and viewing any of the previous inspection reports stored in the office database using spatial queries based on his location and orientation. The spatial database of the bridge and the surrounding environment, and the tracking devices attached to the inspector, make it possible to locate structural elements and detected problems and provide navigation guidance to these objects. In addition, all newly collected information is tagged in space. For example, using a pointing device equipped with a laser range finder, the inspector can point at the location of cracks on the bridge. The exact location can be calculated based on the location and direction of the pointer, the distance to the crack, and the registered 3D model of the bridge.

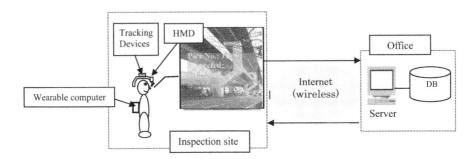

Figure 2.9 Concept of LBC-Infra (Hammad et al., 2004a)

2.5 GEOGRAPHIC INFORMATION SYSTEMS

A GIS is used for managing spatial databases, visualization, and spatial modeling and analysis. Current vector GISs can be categorized into one or more of the following: 2D GISs, 2.5D GISs, or 3D GISs (Figure 2.10). 2D GIS databases contain only the X and Y coordinates of the objects stored in them (points, lines, and polygons). When a GIS database contains the Z coordinate as an attribute of the planar points, the GIS is considered to be a 2.5D GIS. Digital Elevation Models (DEMs) are examples of 2.5D GIS models that can be represented using contour lines or a Triangulated Irregular Network (TIN). 3D GIS databases contain 3D data structure representing both the geometry and topology of the 3D shapes, and allowing 3D spatial analysis.

     (a) 2D GIS   (b) 2.5D GIS   (c) 3D GIS

Figure 2.10 Examples of different GIS categories

A spatial database must contain two types of information about the represented objects: geometric data and topological data. Geometric data contain information about the shape of the objects, whereas topological data include the mathematically explicit rules defining the connectivity between spatial objects (Laurini and Thompson, 1992). Through such topological models, GISs can answer spatial queries about infrastructure objects. Researchers from the GIS, computer graphics, and CAD communities have been investigating spatial data structures and models that can be used as the base of 3D GISs for the past several years (ESRI, 2006).

2.6 GLOBAL POSITIONING SYSTEM (GPS)

GPS is one positioning technology which is available anywhere within certain conditions and it measures the horizontal and vertical positions of the receiver from the GPS satellites. Because of this availability and the relatively good accuracy and low cost of GPS, it is widely used for mobile mapping and other data collection tasks. The GPS consists of 24 earth-orbiting satellites so that it can guarantee that there are at least 4 of them above the horizon for any point on earth at any time (Figure 2.11). The factors that affect GPS accuracy include ionospheric and tropospheric distortion of the radio signals from the satellites, orbital alignment and clock errors of the satellites, and signal multi-path errors (reflections and bouncing of the signal near buildings). In addition, GPS is easily blocked in urban areas, near hills, or under highway bridges. The accuracy of a position is also a function of the geometry of the GPS constellation visible at that moment in time, i.e., when the visible satellites are well separated in the sky, GPS receivers compute positions more accurately. One method to increase the accuracy of GPS is by using Differential GPS (DGPS). DGPS is based on correcting the effects of the pseudo-range errors caused by the ionosphere, troposphere, and satellite orbital and clock errors by placing a GPS receiver at a precisely known location. The pseudo-range errors are considered common to all GPS receivers within some range. Multi-path errors and receiver noise differ from one GPS receiver to another and cannot be removed using differential corrections. DGPS has a typical 3D accuracy of better than 3 m and an update rate of 0.1-1 Hz. The DGPS corrections can be sent to the mobile GPS receivers in real time, or added later by post-processing of the collected data. Real-time kinematic GPS receivers with carrier-phase ambiguity resolution can achieve accuracies better than 10 cm (Kaplan, 1996).

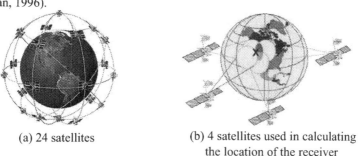

(a) 24 satellites   (b) 4 satellites used in calculating
the location of the receiver

Figure 2.11 Satellites used in GPS

2.7 USABILITY STUDY OF VIRTUAL ENVIRONMENTS

Usability can be broadly defined as *ease of use* and *usefulness* including such quantifiable characteristics such as learnability, speed and accuracy of user task performance, user error rate, and subjective user satisfaction (Hix et al., 2002). Traditional Graphical User Interface (GUI) usability relies on heavily researched and proven tools and heuristics for ensuring effectiveness, efficiency, and satisfaction (Seffah and Donyaee, 2005). Despite intense and widespread research in both VR and usability, until recently there were very few examples of research about the usability of VR methods.

The usability of VE systems is just beginning to receive the attention needed for identifying a taxonomy of VE-specific usability attributes (Gabbard and Hix, 1997; Kalawsky, 1999). Traditional evaluation techniques (Nielsen, 1993) do apply to VEs.

However, they may not be comprehensive enough to characterize usability attributes specific to VEs. Most VEs user interfaces are fundamentally different from traditional GUIs, with unique input/output (I/O) devices, perspectives, and physiological interactions. Thus, when developers and usability practitioners attempt to apply traditional usability engineering methods to the evaluation of VE systems, they find few methods, that are particularly well suited to these environments (Hix and Gabbard, 2002). Subsequently, very few principles for the design of VE user interfaces exist, of which none are experimentally validated (Tromp, 2003). While limited work on VE usability has been conducted to date, there are early works that have attempted to improve VEs from users' perspective by integrating a systematic approach to VE development and usability evaluation (Bowman, 1999; Gabbard and Hix, 1997; Hix and Gabbard, 2002; Kalawsky, 1999; Kaur, 1999; McCauley-Bell, 2002; Sadowski and Stanney, 2002; Stanney, 2003). Stanney et al. (2002) represent the limitations of traditional usability methods for assessing VEs as follows:

(1) Traditional *point-and-click* interactions are not representative of the multi-dimensional object selection and manipulation characteristics of 3D spaces.

(2) Quality of multimodal system output (e.g., visual, auditory and haptic) is not comprehensively addressed by traditional evaluation techniques.

(3) Means of assessing sense of presence and after-effects have not been incorporated into traditional usability methods.

(4) Traditional performance measurements (i.e., time/accuracy) do not comprehensively characterize VE system interaction.

(5) Traditional single-user-task-based assessment methods do not consider VE system characteristics in which two or more users interact in the same environment.

Stanny et al. (2002) classify VE system usability into two main categories: VE system interface (i.e., software, hardware, and overall man-machine interaction design) and VE user interface (i.e., physiological, psychological, and psychosocial) considerations (Figure 2.12). More details on the interfaces of VE systems are given in the following section.

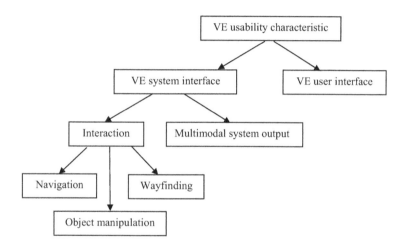

Figure 2.12 Usability criteria (Stanny et al., 2002)

2.7.1 Virtual environment systems interface considerations

The users of VE systems should be able to effectively interact with optimized multimodal system output. In a VE system, the computer displays and controls are configured to involve users in an environment containing 3D objects (Durlach and Mavor, 1995). Each virtual object has a location and orientation independent of users' viewpoints. Users can interact with these objects in real time using a variety of motor and display output channels. Consequently, interaction design needs to sufficiently match how people use a system to perform their tasks. This matching is characterized by interaction techniques that combine an understanding of human capabilities (i.e., communication, cognitive and perceptual skills) with computer Input/Output (I/O) devices and machine perception and reasoning (Turk and Robertson, 2000). In addition, a system should perceive relevant human communication channels and generate output that is naturally understood by users. Moreover, a usable VE system needs to incorporate a flexible underlying model or simulation allowing users to control the VE and interact with it in a natural, task-supported manner (Gabbard and Hix, 1997; Kalawsky, 1999; Kaur, 1999). The usability of any system is thus influenced by VE system components as they impact users' ability to complete tasks for which the system is designed to support (Bowman, 1999; Gabbard et al., 1999). Bowman et al. (2000) noted that a distinction must be made when referring to input devices versus interaction techniques because input devices are simply the physical tools utilized to implement various interaction techniques. Using this distinction, a VE system interface can be characterized by the interaction techniques (as user inputs are inherently included) and the mutli-modal system outputs (conveying interaction feedback and other system information) employed in a system design.

2.7.2 Interaction in virtual environments

Interactivity is the most critical attribute of VEs, as it clearly defines the difference between VR and other 3D modeling systems. The interaction means that any action on the part of the user results in a change in the virtual environment. A simple interaction is

26

selecting one object to initiate changes in the scene. Interaction can also include navigation where a user moves through a VE while it updates and changes according to the user's movements.

Bowman (1999) proposed that any interaction within a VE falls into three general categories: (1) Travel (i.e., movement of user's viewpoint from place to place); (2) Selection (i.e., targeting virtual objects within an environment); and (3) Manipulation (i.e., setting the position and/or orientation of virtual objects). Herndon et al. (1994) classified basic user interactions in a VE to be navigation, viewpoint control, and object interaction. Other studies have focused specifically on navigation when assessing interaction behavior (Benyon and Hook., 1997). Still others have defined interaction to be the general look, feel and behavior as users interact with an application (Hix and Gabbard , 2002; Gabbard et al., 1999). These latter researchers consider that user interaction components consist of all icons, text, graphics, audio, video, and devices through which users communicate with and move within a VE. To affect such basic human behaviors associated with communication within VEs, users must navigate, approach/orient, and execute (Sutcliffe and Kaur, 2000).

According to the above descriptions and the notion that focusing on processes involved in interaction can ensure users' abilities to successfully complete general interaction tasks (Stanny, 2003), usability criteria associated with interaction have been classified as: wayfinding (i.e., locating and orienting oneself in a VE); navigation (i.e., moving from one location to another in a VE); and object selection and manipulation (i.e., targeting objects within a VE to reposition, reorient and/or query). The following sections provide a short review of the research related to these interactions.

2.7.3 Wayfinding in virtual environments

VE navigation involves how users manipulate their viewpoint to move from place to place within the environment (Bowman, 1999; Hix and Gabbard, 2002; Gabbard et al., 1999; Kaur et al., 1999). Darken and Sibert (1996a) identified the cognitive component of VE navigation as wayfinding, or the ability to maintain knowledge of one's location and orientation while moving throughout a designed space. Users have been found to experience difficulty maintaining knowledge of their location and orientation while traversing a VE (Chen and Stanney, 1999; Darken and Sibert, 1996b). Accordingly, users may devote much of their interaction time trying to figure out the spatial layout of a VE and this can detract them from focusing on task objectives. If there is insufficient or inappropriate information provided about the spatial structure of a VE or identity and location of target objects, then users are likely to have difficulties locating their current and/or desired destinations (Kaur et al., 1999). The design of a VE must, therefore, include appropriate visual cues and navigational aids, such as a map, to facilitate users' acquisition of spatial knowledge. Without proper design of the navigational space and availability of tools to aid in exploring this space, the overall usability of a VE system will suffer resulting in ineffective and inefficient task performance. The concept of travel (i.e., users moving from place to place) must also be addressed. In wayfinding in VE systems, users must be able to effectively move in the environment to obtain different views and acquire an accurate "mental map" of their surroundings (Bowman, 1999; Chen et al., 1999).

A number of researchers are beginning to identify specific usability issues that must be considered to ensure VE wayfinding is effectively supported. For instance, Furnas (1997) discussed some basic requirements for wayfinding in abstract information spaces (e.g., users should be able to use the current view to plan the shortest path to a target). Charitos and Rutherford (1996) defined the general requirements for designing spatial structures (e.g., paths should have a clear structure and start/end points; landmarks should be easily identifiable and recognizable with a prominent spatial location). Kaur et al. (1999) cited a variety of recent research findings and proposed new guidelines regarding wayfinding in both VEs and information spaces in general.

2.7.4 Navigation in virtual environments

Navigation is the most basic and common type of interaction within a VE (Bowman, 1999). For most VE users, navigation is what is necessary to allow users to move into a position to perform required tasks. Navigational techniques should be easy to use and not cognitively cumbersome (Stanney et al., 2003). Unfortunately, current VE navigational techniques have not always met these criteria. Researchers are beginning to focus on identifying general issues surrounding usability of VE navigational techniques (Herndon et al., 1994). Other researchers proposed various metaphors for viewpoint motion and control in 3D environments such as *flying, eyeball-in-hand, world-in-miniature, possession and rubberneck navigation* and *speed-coupled flying with orbiting* (Pausch et al., 1995; Ware and Jessome, 1988; Tan et al., 2000). Other studies have evaluated non-immersive 3D navigational techniques (Strommen, 1994; Ware and Slipp, 1991). Given the copious techniques available to support travel, it is difficult to determine exactly which components of these techniques are significant in improving or lessening navigational performance (Bowman, 1999) especially when considering navigation in real environments.

It is obvious that users should be able to interact with and control their movement throughout a VE in a natural and flexible way to support their task (Gabbard and Hix, 1997; Kalawsky, 1999; Kaur et al., 1999). Tan et al. (2000) introduced a taxonomy of navigation methods including *object manipulation and ghost copy, inverse fog/scaling and ephemeral world compression, possession and rubberneck navigation,* and *speed-coupled flying combined with orbiting.* The last method takes the user viewpoint up when the speed of navigation is high and brings it down when the speed is low (Figure 2.13). Tan et al. (2000) showed the efficiency of this navigation method based on usability studies. However, this method could be efficient in game applications but it is not suitable for engineering applications requiring navigation in UVEs because it does not provide a natural way of navigation.

(a) Local view of scene while moving slowly

(b) Overview of scene while moving fast

Figure 2.13 Speed-coupled flying with orbiting (Tan et al., 2000)

2.7.5 Object selection and manipulation in virtual environments

Object selection and manipulation can be defined as the process of indicating virtual objects within an environment to reposition, reorient, or query them (Bowman, 1999; Gabbard and Hix, 1997). Object selection involves users designating one or more virtual objects for some purpose (e.g., deleting an object, invoking a command, changing system-state). This often is followed by subsequent manipulation of specified objects. The selection and manipulation methods incorporated into a VE design have a profound impact on the quality (i.e., efficiency and effectiveness) of interaction. Basic interaction issues (e.g., not being aware of which objects are active) appear to be more prevalent in VEs than traditional direct manipulation interfaces. Because unlike the traditional interfaces, standards for this action is not established in VEs, such as how to define active objects (Stanny, 2002). Furthermore, the added spatial dimensions intrinsic to VEs may place more demand on object manipulation precision. Therefore, the tracking methods are important in navigation and object manipulation.

2.8 SUMMARY AND CONCLUSIONS

In this chapter, the literature about the current situation of VR and its applications in urban environment has been reviewed. There are different approaches for creating the 3D city models. However, a lot of time and money have to be invested to design the model and to integrate the necessary data. In addition, these models focus only on the exterior shapes of buildings and do not include all the details necessary in civil engineering applications. Furthermore, the usability testing of interaction and navigation in VEs are reviewed. Little work has been done about providing interaction and navigation support in engineering UVEs and their usability testing.

Based on the literature review, the main focus of the research will be developing a practical method for creating UVEs for civil engineering applications and improving the efficiency of interaction and navigation in these models. In addition, a new approach for usability testing of the UVEs is necessary.

CHAPTER 3   FRAMEWORK FOR VIRTUAL REALITY
APPLICATIONS IN CIVIL ENGINEERING

3.1 INTRODUCTION

As introduced in Chapter 2, 3D city models have been useful for many urban applications such as urban planning and construction management. However, in order to realize a 3D urban model, an enormous amount of time and money has to be invested to design the model and to integrate the necessary data. Large-scale UVEs can be created using commercial GIS software. However, these models focus only on the exterior shapes of buildings and do not include all the details necessary in civil engineering applications (e.g., design and scheduling information).

In this chapter, we introduce a data integration method for creating a system to generate UVEs automatically by integrating GIS maps, CAD models, images of building facades, and databases of the cost, scheduling, and other data generated during the lifecycle of buildings. In addition, we will introduce a framework for using these models to suit the requirements of mobile infrastructure management systems. This framework is called *Location-Based Computing for Infrastructure field tasks* (LBC-Infra). LBC-Infra can be used by field workers to interact with geo-referenced infrastructure models and automatically retrieve the necessary information in real time based on their location and orientation, and the task context (Hammad et al., 2004a).

In Chapter 4, we will discuses about interaction component of LBC-Infra and how to improve the interactivity and efficiency of navigation. Furthermore, in Chapter 5, Concordia University downtown campus will be used in the case studies to demonstrate the proposed framework. The 3D virtual model of the university campus will be developed and the necessary interaction and navigation components will be added to the model. The navigation and wayfinding in this model will be tested based on user studies.

3.2 FRAMEWORK FOR VR APPLICATIONS IN CIVIL ENGINEERING

The proposed framework is based on a collection of high-level reusable software objects and their relationships. The generic structure of this framework embodies the general functionalities of MR applications in civil engineering so that the structure can be extended and customized to create more specific applications, e.g., a building inspection application or a construction progress monitoring application. Because the framework integrates several evolving technologies, its implementation should follow an open and extensible design so that the application development can adapt to new requirements and new technologies while reducing the time and cost of the development. The Model-View-Controller (MVC) software development model (Potel, 1996) has been selected as the basic framework of the system because of its simplicity and flexibility in manipulating complex and dynamic data structures requiring diverse representations. The MVC model has three main high-level objects: *Model*, *View*, and *Controller*. The *Model* represents the data underlying the application that are accessible only through the *View* object. The *Model* of the system has three basic databases for managing the spatial data, data about the attributes of buildings (e.g., inspection data), and data about field tasks (e.g., information about inspection tasks, their order, and devices and methods used for performing them). The *View* object accesses the data from the *Model* database and specifies how these data are presented to the user, e.g., the information about the components of a building can be used to create the 3D model of the building. The

Controller determines how user interactions with the *View*, in the form of events, cause the data in the *Model* to change, e.g., adding a defect to a structural element causes the value of the "inspection" attribute of the selected element to be set to the specific defect type in the attribute database. The *Model* closes the loop by notifying the *View* to update itself so that it reflects the changes that occurred in the data. Using the MVC model, new methods of interaction can easily be introduced to the system by developing new *View* objects.

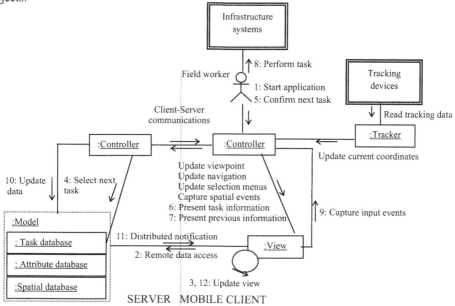

Figure 3.1 Collaboration diagram between the entities of LBC-Infra (Hammad et al., 2004a)

Using the MVC model in distributed client-server applications involves deciding which parts of the model are implemented, in whole or in part, on the client or the server. The *Model* represents typical server-side functionality while the *View* represents typical client-side functionality. The *Controller* can be partitioned between the client and the

35

server; most of the processing can be on the client side in a fat client or on the server side with only a simple GUI application on a thin client. In more elaborate architectures, various partitioning of the functionalities between the client and the server are possible, e.g., the client may have a proxy of the *Model* to lower the demand for communication with the server. In addition to the *Model, View* and *Controller,* the system has a *Tracker* object that handles location-related functionalities as explained in the following paragraphs.

Figure 3.1 shows the relationship between the high-level objects of the system using a Unified Modeling Language (UML) collaboration diagram, where objects interact with each other by sending messages. The numbers in the diagram refer to the order of execution of the messages. Messages that are not numbered are threads that are executed at the beginning of the application and run continuously and concurrently with other messages, or they are event-driven messages that may occur at any time. A *Field Worker* starts interacting with the system by sending a message (*start application*) to the *Controller* (message 1). As part of the initialization of the system, the *View* creates its contents (message 2) by retrieving the necessary information from the databases of the Model that reside on a remote server. It then updates itself (message 3). Once the application is initialized, the *Tracker* starts continuously reading the location and orientation measurements of the *Field Worker* from the tracking devices and updating the *Controller* about the current coordinates of the *Field Worker*. The *Controller* retrieves the information about the next task to be performed from the *Task Database*, which contains a plan defining the tasks (message 4). However, the *Field Worker* has the freedom to confirm this selection or override it by selecting another task if necessary (message 5).

Based on the changing coordinates of the *Field Worker*, the *Controller* updates the viewpoint of the *View*, filters the contents of the selection menus, and updates the navigation information in the direction of the best location and orientation to perform the present task. These updating and filtering steps ensure that the information presented coincides with what the *Field Worker* can see in the real scene. The navigation is performed by presenting visual or audible guidance. As the *Field Worker* follows the navigation guidance towards the new location, the *Controller* provides him or her with information about the task (message 6). Another function of the *Controller* is to capture spatial events, such as the proximity to an element of interest. This flexibility allows switching the order of the tasks to inspect an element planned for inspection in a subsequent task. As in message 5 above, the *Field Worker* can accept or reject this change in the task order. Once the *Field Worker* is at the right position and has the right orientation to perform the task at hand, the *Controller* presents the worker with previously collected information (if available) that may help in performing the present task (message 7). At this point, the *Field Worker* performs the task, e.g., the collecting of inspection data visually or by using some devices (message 8). He or she can input the collected data by interacting with the *View*, e.g., by clicking on an element displayed within the *View* for which data have to be updated. The input events are captured by the *View* (message 9) and used by the *Controller* to update the data in the *Model* (message 10). Finally, the changes in the databases are channeled to the *View* by distributed notification (message 11) so that the *View* can update itself (message 12). Client-server communications can happen whenever needed depending on the nature of the client (thin or fat).

3.3 DATA INTEGRATION FOR CREATING THE VIRTUAL MODEL

One of the important issues in developing VR applications for civil engineering is to use a systematic method in creating the virtual model. Large-scale virtual models of urban environments (e.g., a university campus) can be created using commercial GIS software. However, these models focus only on the exterior shapes of buildings and do not include all the details necessary in civil engineering applications (e.g., design and scheduling information).

The proposed data integration method for creating the virtual building model is based on synthesizing information from GIS maps, CAD models, images of building facades, and databases of the cost, scheduling, and other data generated during the lifecycle of buildings. Two representations are created for each building with different Levels of Details (LoD): one is for the exterior of a building (LoD$_e$) and the other is for the detailed 3D model of the interior of the same building (LoD$_i$). Images are applied as texture mapping on the facades of the models of the buildings to make the buildings more realistic and easier to recognize.

In large-scale virtual models, several buildings should be added to complete the virtual model of the area surrounding the buildings considered in the system, but the details of these buildings are not needed. In this case, only the exterior shapes of the buildings (LoD$_e$) are created. The steps used in this integration method are briefly explained (Figure 3.2).

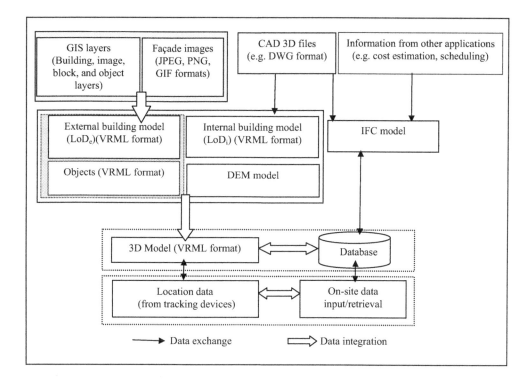

Figure 3.2 Data exchange and integration model

3.3.1 Creating GIS layers

(1) For the exterior representation of a building (LoD$_e$), the footprint of the building is constructed and added to a building layer in the GIS (polygon layer). In case a building has changes in its perimeter with respect to the height, additional polygons are added to capture these changes. Figure 3.3(a) shows an example of the CAD data used to create the footprints of one building. The base level and the height of each polygon are added as attributes of that polygon.

(2) The images of the facades are collected and processed to create orthogonal images. These images are added to the surfaces of the models of the buildings by applying

texture mapping techniques. To this end, the locations of the surfaces corresponding to these images are represented by an image layer in the GIS (line layer). Figure 3.3(d) shows an example of the image layer for one building. The base level, height, and file name of each image are added as attributes of that image.

(3) A block layer is also added in the GIS (polygon layer) to represent pedestrian areas surrounding buildings as shown in Figure 3.3(f). A nominal height is added as an attribute to all polygons to represent the height of the pedestrian areas above the pavement level of the roads.

(4) Other objects of interest, e.g., traffic lights, street lights and street furniture, may be added to the 3D model as part of the scope of the virtual model. These objects are added to an object layer (point layer) in the GIS as shown in Figure 3.3(e). For simplicity, it can be assumed that all objects of the same type have the same shape, and therefore a standard library of 3D shapes can be created and used to generate a specific object (Figure 3.4).

3.3.2 Translation to virtual reality model

In order to translate the GIS layers to VR model, the layers mentioned above (building, image, block and object layers) are used to automatically extrude the 2D shapes into 3D shapes, to add the texture mapping, and to insert the 3D objects into the virtual 3D model. An example of the result of this synthesis is shown in Figure 3.3(g).

3.3.3 Linking the model to database

In addition to the geometry, each element of a building (or other objects) in the virtual model is linked through a unique identification number to a database where the attributes related to that building are saved. The database can serve as a lifecycle database for cost, scheduling and other information of maintenance and inspection activities. The Industry Foundation Classes (IFC) can be used as the interoperability standard to export/import data to/from engineering applications (Seren and Karstila, 2001). When using the application on site to collect data, tracking devices can be used to add the location of the collected data to the 3D model.

3.3.4 Adding interior models of buildings

For the interior representation of a building (LoD$_i$), the 3D CAD model of the building is prepared and translated to a suitable format, such as the Virtual Reality Modeling Language (VRML). Figure 3.4 shows an example of a floor in a building. In order to locate the model of the building in the virtual 3D model, two points are identified in both models (e.g., two points on one edge of the building), and their coordinates are used to calculate the transformation matrix, including rotation, scaling and translation, from the local coordinate system of the CAD model to the global coordinate system of the virtual model. Another issue to be considered is the orientation of the coordinate axes used in different visualization software. For example, the axis in the height direction is considered as the Y-axis in Java 3D (the 3D API of Java language) and as the Z-axis in 3D Studio Max.

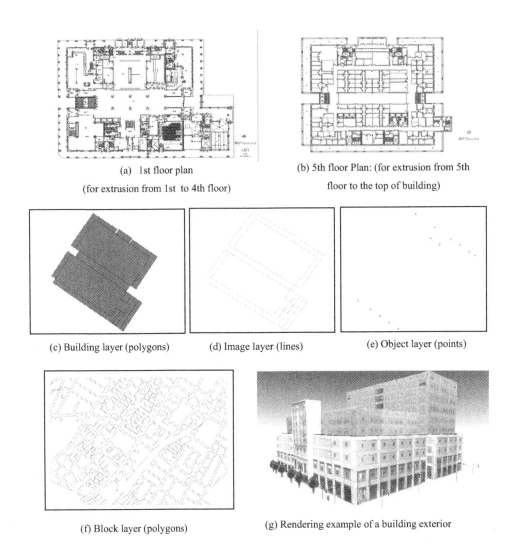

(a) 1st floor plan
(for extrusion from 1st to 4th floor)

(b) 5th floor Plan: (for extrusion from 5th floor to the top of building)

(c) Building layer (polygons)

(d) Image layer (lines)

(e) Object layer (points)

(f) Block layer (polygons)

(g) Rendering example of a building exterior

Figure 3.3 Example of data used in creating 3D models

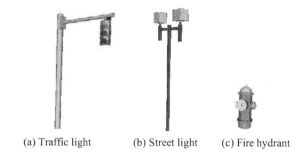

(a) Traffic light   (b) Street light   (c) Fire hydrant

Figure 3.4 Examples of objects from the 3D object library

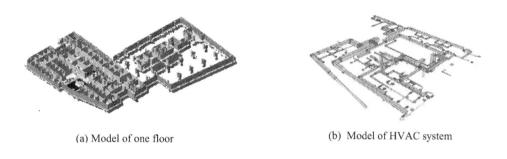

(a) Model of one floor   (b) Model of HVAC system

Figure 3.5 Examples of detailed 3D CAD models

3.4 TRACKING METHODS

Three tracking methods are used in LBC-Infra: (1) Tracking the location of the user in the VE; (2) Tracking the location of the user in the real world in indoor applications; and (3) Tracking the location of the user in the real world in outdoor applications.

Tracking the location of the user in the VE is done by tracking the viewpoint. Several sensor-based positioning systems can be used indoors, such as video, electromagnetic, infrared or ultrasonic systems (Karimi and Hammad, 2004). Video tracking is used to track visual markers by means of a video camera (Kato, 2000). The GPS can be used for outdoor tracking. Figures 3.6 (a) and (b) show the video-based and GPS tracking methods

investigated in this work for indoor and outdoor tracking, respectively. In both cases, the

user is equipped with a tablet PC, an electronic stylus, and a digital camera fixed on the

hardhat for recording images and video clips. However, in figure 3.6(a), the video camera

has an additional role of recognizing markers attached to the walls at known locations.

After detection of the location of the user using tracking devices, the location of the

viewpoint in the 3D model will be adjusted with the real world. For example, the

inspector will see on the screen of the tablet PC the same part of the facility that he/she is

inspecting. In order to add a damage that has been found to the 3D model, the inspector

can click on the element to directly add a damage, which is represented by a 3D shape on

the surface of the inspected element.

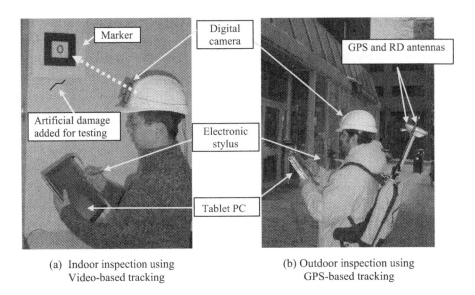

(a) Indoor inspection using
Video-based tracking

(b) Outdoor inspection using
GPS-based tracking

Figure 3.6 Examples of inspection for FMIS (Mozaffari et al., 2005)

3.5 SUMMERY AND CONCLUSIONS

In his chapter, a practical data integration method for creating UVEs is proposed. The following conclusions about the proposed approach can be stated: (1) The UVEs are discussed within a larger framework called LBC-Infra. The generic structure of this framework embodies the general functionalities of VR applications in civil engineering so that the structure can be extended and customized to create more specific applications, e.g., a building or bridge inspection application or a construction progress monitoring application; (2) A data integration method is proposed for creating the UVEs by synthesizing information from GIS maps, CAD models, images of building facades, and databases of the cost, scheduling, and other data generated during the lifecycle of buildings; and (3) Three tracking methods that are provided in the framework are introduced.

CHAPTER 4   INTERCTION IN URBAN VIRTUAL ENVIRONMENTS AND USABILITY METRICS

4.1 INTRODUCTION

In Chapter 3 a systematic approach for creating the integrated 3D model of an UVE is proposed and then this model is discussed within a larger framework (LBC-infra) for VR applications in civil engineering. In this chapter, the interaction components of this framework are introduced. We will discuss the design of some of these interaction methods that will improve the performance of the applications based on the framework.

LBC-Infra integrates 3D models, tracking technologies, mobile computing and distributed wireless communication in one framework. This integration will result in the following advantages: (1) Visualizing different types of data; (2) Providing a user-friendly interface which can facilitate interaction with 3D models and reduce data input errors; (3) Facilitating data sharing; and (4) Improving the efficiency of database management. 3D visualization can be understood more quickly and completely than the traditional management tools. In large-scale infrastructure projects, GISs are inevitably needed for generating information that relates to locations. Spatial interactions would not be understood fully if they were not linked to geographical locations as perceived in the real world. Therefore, the 3D models should be located on a 3D map.

As discussed in Chapter 2, one of the important issues in realizing VR applications in civil engineering is the difficulty of navigation in large UVEs. To accomplish their tasks in these environments, the users need to be able to navigate either to accomplish specific tasks or to become more familiar with the UVEs. Properly designed user interfaces for navigation can make that experience successful and enjoyable. A large number of studies have been carried out on the general principles of 3D navigation and wayfinding in virtual environments. However, little work has been done about providing navigation support in engineering VR applications involving large scale urban environments.

This chapter has the following objectives: (1) To investigate the interaction and navigation components of LBC-Infra; (2) To investigate a taxonomy for navigation methods and support tools for engineering applications in UVEs; and (3) To propose a GIS-based evaluation technique for usability testing of navigation in UVEs.

4.2 INTERACTION COMPONENTS OF LBC-INFRA

LBC-Infra has the following main interaction components: visualization and feedback, control, access, navigation, manipulation, and collaboration. The following interaction patterns for each component are examples that have been identified based on common tasks that field workers usually perform and the type of information they collect.

(1) Displaying graphical details: LBC-Infra displays to the field worker structural details retrieved from previous inspection reports. This can happen in a proactive way based on spatial events. For example, once a cracked element is within an inspector field-of-view, the system displays the cracks on that element discovered during previous inspections. This will help focus the inspector's attention on specific locations. The user of the system can control the Levels of Details (LODs) of representing objects depending on his or her needs;

(2) *Displaying non-graphical information and instructions*: The user interface can provide links to documents related to the project, such as reports, regulations and specifications. In addition, LBC-Infra allows for displaying context-sensitive

instructions on the steps involved in a specific task, such as instructions about the method of checking new cracks, and measuring crack size and crack propagation.

(3) *Control*: LBC-Infra interprets the user input differently depending on the selected feature and the context. For example, clicking the pointing device can result in selecting a menu option or in picking an object from the 3D virtual world depending on where the user clicked.

(4) *Access*: Accessing data can be achieved based on location tracking. The user just needs to walk towards an object of interest and then click on it or stand in front of it for a short period of time. The application tracks his/her location and updates the viewpoint accordingly. Based on the location, relevant data are retrieved from the database and displayed

(5) *Navigation*: As an extension to conventional navigation systems based on 2D maps, LBC-Infra can also present navigation information in 3D VE. Within a specific field task, the system can guide a field worker by providing him/her with navigation information and focusing his or her attention on the next element to be inspected.

(6) *Manipulation*: Once an object is found, the inspector can add inspected defects by directly editing their shapes in VE and manipulating them using a pointing device.

(7) *Collaboration*: LBC-Infra facilitates wireless communications among a team of field workers, geographically separated in the project site, by establishing a common spatial reference about the site of the project. In some cases, the field workers may collaborate with an expert engineer stationed at the office who monitors the same scene generated by the mobile unit in the field.

In this book, we focus on the visualization and navigation components that are suitable and practical for engineering applications in UVEs.

4.3 NAVIGATION TAXONOMY

Based on our experience with similar systems and our review of previous research, we have defined the taxonomy of navigation methods and tools for civil engineering applications (Figure 4.1).

The navigation tasks are classified into three categories: (1) Naive search, in which the navigator has no a priori knowledge about the location of the target; (2) Primed search, in which the navigator knows the location of the target; and (3) Exploration: Wayfinding tasks in which there is no specific target. The navigation in the 3D model can be realized by using tracking devices, such as a GPS tracking or video-based tracking, or pointing devices, such as a digital stylus.

Four types of navigation behavior can be used: drive, fly, orbit, and examine behavior (more details are given in Section 4.4.2). The navigation speed can be controlled manually or automatically. In the manual mode, the user can change the speed of navigation using a slider. In the automatic mode, the speed of navigation can be changed according to the distance from the present location to the target. The navigation slows down when the user approaches the target and speeds up when he/she moves away from it. This method is more suitable for civil engineering applications than other more sophisticated approaches, such as the *speed-coupled flying with orbiting technique*. Navigation supports can be grouped into six categories:

(1) *Cues:* Four types of cues are considered: landmarks, building names, road names and the depth cue. Figure 4.2(a) shows examples of cues that are used in our

applications. The Digital Elevation Model (DEM) is integrated with the VR model of the built environment to display natural landmarks (e.g., mountains and rivers). Furthermore, the distinctiveness of some buildings is enhanced by adding texture mapping in order to increase the memorability of those buildings and their locations. In addition, building and road names can be shown on top of the buildings and roads using billboard behavior (more details are given in Section 4.4.2.2). To improve the realism of the VE, the stereoscopic effect can be used, which provides a depth cue for the user; as a result, the distances and dimensions can be better perceived.

(2) *Animated arrows:* Navigation guidance can be provided with an animated 3D arrow showing the user the path to the object of interest (Figure 4.2(b)). The arrow moves from the current users' location towards the object. If the user cannot take a straight path to the object for any reason, such as the presence of obstacles, he/she can take a different way without losing the object because the arrow will be automatically updated to go from his/her new location to the object (more details are given in Section 4.4.2.2).

(3) *2D maps and floor plans:* Electronic maps are good navigation aids that help users to explore and learn about the space and to orient themselves. Figure 4.2(c) shows an electronic map of the area beside the 3D model with the present location of the user. In addition, 2D floor plans can be displayed in a different window beside the 3D model or embedded in it to ease the wayfinding process inside a building (Figures 4.2(d, e)).

(4) *Constrained movement:* Constraining the movement of the user in the 3D space can facilitate the navigation by preventing the user from going below the ground, walking

through walls or going out of the UVE boundaries. The constrained movement is provided in LBC-Infra by applying terrain following, collision avoidance and space boundaries constraints. Terrain following is particularly important in navigation within terrestrial style environments. Collision detection is important in navigation within environments which contain a large number of objects. Space boundaries keep the user inside the UVE.

(5) *Viewpoints*: Static viewpoints provide a link to a familiar place with a known position. If the user gets lost, the viewpoints can help him/her to get back to a safe and familiar point. There are two types of static viewpoints: predefined viewpoints that are already defined and user-defined viewpoints that are created by the user during the navigation in the VE. The tours provide a set of viewpoints (a path) through which the user can cycle to visit the UVE in a certain order (more details are given in Section 4.4.2.3).

(6) *Tree structure:* The tree structure is another method for managing the interaction with the components of the VE. It acts as an index by showing the hierarchy of objects in the UVE. Users can look for an element in the tree such as a building (or a floor within a building) and select it. Consequently, the selected building will be highlighted, and the viewpoint will change to navigate to it (Figure 4.2 (f)).

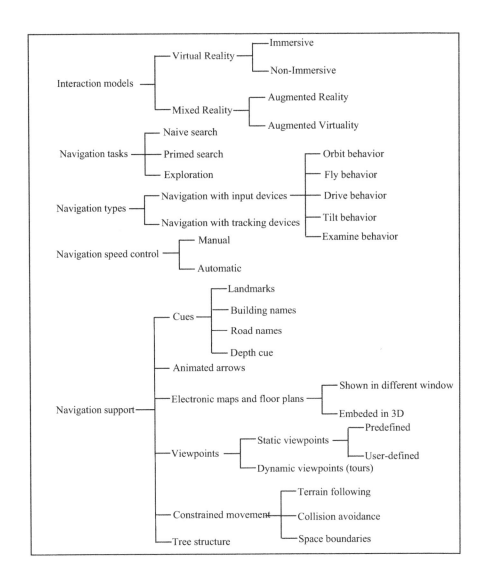

Figure 4.1 Taxonomy of interaction models and navigation

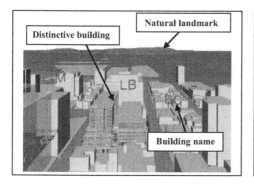

(a) Examples of cues

(b) Animated arrow pointing to the objects of interest from the current location of the user

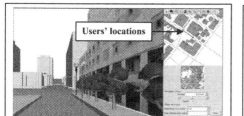

(c) 2D map of the area showing current location of the user

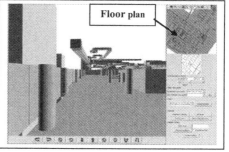

(d) Floor plan displayed in different window

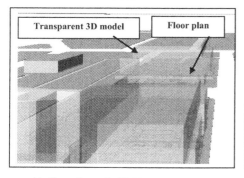

(e) Floor plan embedded in the 3D model

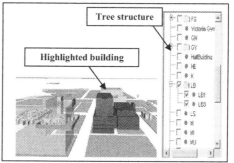

(f) Highlighting a building using the tree structure

Figure 4.2 Examples of navigation supports

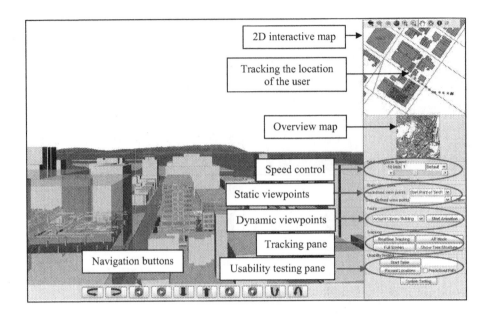

Figure 4.3 General structure of user interface in LBC-Infra

4.4 INTERACTION WITH THE 3D MODEL

4.4.1 User interface design

In order to show the 3D model on the small screen of a tablet PC used in mobile situations in civil engineering applications, a simple but efficient GUI should be designed considering the navigation and interaction functions of LBC-Infra. The GUI should facilitate the interaction with 2D GIS and 3D virtual environments in real time and provide consistent feedback to the user. The main area of the interface is used to show the 3D browser (Figure 4.3). Navigation and interaction can be facilitated by the tree structure of the campus, speed control interface, predefined static and dynamic viewpoints and 2D interactive map of the area.

As discussed in Chapter 3, tracking is one of the technologies integrated in LBC-Infra to satisfy the mobile location-based requirements of LBC-Infra. The GUI should provide the user with the options to change the tracking mode in the real world or in the VE. For example, tracking of the user in the real world can be done using video-based tracking for indoor tracking (e.g., data collection for FMS inside a building) or the GPS for outdoor tracking (e.g., bridge inspection). The tracking of the user in the VE can also be done by tracking the location of the viewpoint. In addition, by default the user current location is shown in the 2D interactive map and the map is scaled at the user location. As shown in Figure 4.4, the current location of the user is zoomed and shown by a red triangle and previous locations are shown by green circles. In addition, the map can be rotated so that the heading matches the direction of movement. The user can disable these functionalities using the GUI.

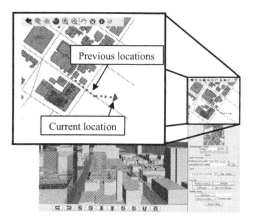

Figure 4.4 2D interactive map

4.4.2 Navigation and picking behaviors in the VEs

A dynamic environment can be represented by a set of objects that have particular behaviors. The interactivity in LBC-Infra is mainly facilitated using behaviors such as navigation and picking behaviors. These behaviors can be *environment-independent* behaviors, which do not consider the current state of other objects in the environment, or *environment-dependent* behaviors, which do consider other objects (Kessler, 2002). The environment-independent behaviors used in LBC-infra include time-based behaviors. The environment-dependent behaviors include event-driven behaviors, which respond to events initiated by users or other objects, and constraint maintenance behaviors, which react to changes of other objects to maintain defined constraints.

4.4.2.1 Time-based behaviors

(1) *Interpolators*

The interpolator behavior provides animations by interpolating among two extreme values (alpha) over time (Figure 4.5). An alpha value of 0.0 generates the interpolator's minimum value, an alpha value of 1.0 generates the interpolator's maximum value, and an alpha value somewhere in between generates a value proportionally in between the minimum and maximum values. The interpolation among a set of predefined positions of the viewpoint during a specific time is used in the framework to create the tours described in Section 4.3.

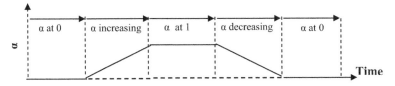

Figure 4.5 Interpolator's generic time-to- α mapping sequence

56

(2) *Transformation behaviors*

The transformations of the viewpoint (rotations and translations) based on time elapse are provided in the framework through the transformation behaviors. The translation of the viewpoint can be to the right, to the left, up, down, forward and backward. The rotation of the viewpoint can be around the local Y axis (Yaw) or around the local X axis (Pitch). These behaviors are activated by pushing the navigation buttons (Figure 4.6). While the button is pressed, the requested transformation will be performed every specific period of time.

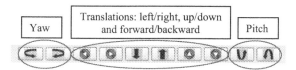

Figure 4.6 Navigation buttons

4.4.2.2 Event-driven behaviors

(1) *Picking behavior*

Picking is the process of selecting shapes in the 3D VE using the 2D coordinates of the picking device. A pick shape is selected as the picking tool. The pick shape could be a ray, segment, cone, or cylinder. The pick shape extends from the viewpoint location, through the picking device location and into the VE. When a pick is requested, pickable shapes that intersect with the pick shape (e.g., pick ray) are computed. The pick returns a list of objects, from which the nearest object can be computed.

Figures 4.7 and 4.8 show the flowchart and an example of the picking behavior, respectively. Each object in the VE (e.g. buildings, bridges and elements of them) has a predefined ID, which is related with the data stored in the database. Upon selection, the object will be highlighted and a query is activated to retrieve the matching information in the database.

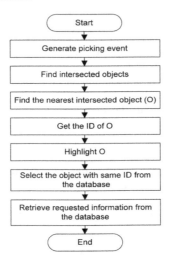

Figure 4.7 Flowchart of picking and highlighting an object

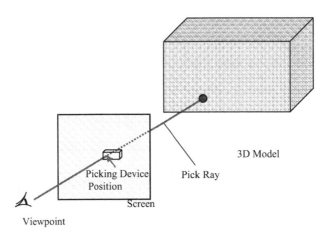

Figure 4.8 Example of picking an object in the VE

58

An application that is facilitated by picking in the framework is adding an object to the VE, such as adding a workspace for a column inside a building. Upon the selection of a column, a workspace around that column is created and the information about the column is displayed such as the dimensions of the column (Figure 4.9(a)). This function is useful for workspace analysis. The purpose of workspace analysis is to investigate the workspace representation and conflict detection in the case of construction and maintenance activities in civil engineering applications.

Another application in the case of inspection of elements of a structure is that the user can mark defects, which are represented by 3D shapes, on the surface of the inspected element (Figure 4.9(b)).

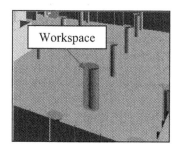
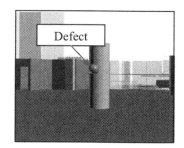

(a) Workspace around a column   (b) Defect on a column

Figure 4.9 Examples of adding an object to the VE

(2) *Navigation behaviors*

Four types of navigation behavior are provided in LBC-Infra: *drive*, *fly*, *orbit* and *walk*. These types of behavior use the pointing device to control the viewpoint motion. Each button on the pointing device generates a different type of motion while the button is pressed. The distance of the pointing device location from the center of the display area controls the speed of motion. As an example of the navigation behavior, the drive behavior allows the user to move to any point in the VE, with pointer controls for translations along the X, Y, and Z axis and rotation around the Y axis (Figure 4.10).

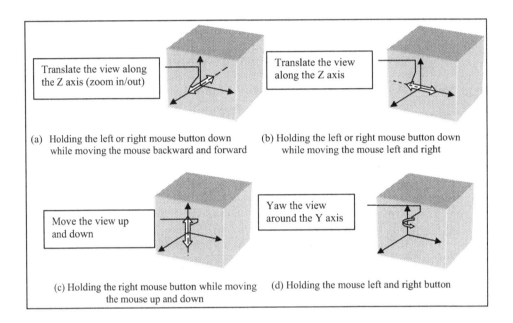

Figure 4.10 Drive behavior

(3) *Automatic speed behavior*

The speed of the navigation in the VE can be adjusted automatically based on the distance from center of the VE. This behavior is activated every time the location of the viewpoint changes and it calculates the distance between the present position of the viewpoint and the center of the VE. Then, the speed of motion is adjusted proportional to this distance.

(4) *Real-time navigation guidance behavior*

This behavior provides real-time navigation guidance in LBC-infra through animated 3D arrows showing the user the path to the object of interest. This behavior is activated when the viewpoint changes.

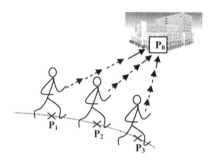

Figure 4.11 Animated arrow pointing to the object of interest

The system gets new locations from the tracking devices in real time and changes the viewpoint accordingly (Figure 4.11). Then, it adds an arrow that moves between the location of the user (e.g., P_1) and the location of the object of interest (P_0). The animation of the arrow is provided by interpolating the arrow over a specific time between two points (the current location the user and the object of interest). The orientation of the arrow follows a vector that goes from the location of the user to P_0.

(5) *Interaction with the 2D map behavior*

Interaction with the 2D map of the area is provided by this behavior. This behavior is activated by changing the location of the viewpoint in the VE or by defining two points on the 2D electronic map (Figure 4.4). By changing the location of the

61

viewpoint in the VE, the present location of the user is shown on the 2D map. In addition, upon clicking on the two points on the 2D map, the viewpoint in the VE will be changed. These points specify the new location and direction of the viewpoint.

(6) *Billboard behavior*

This behavior adjusts the orientation of the road and building names that are used as visual cues in LBC-Infra such as they always face the user.

4.4.2.3 Constraint maintenance

(1) *Collision avoidance and terrain following behaviors*

The basic process of collision avoidance and terrain following is detecting where the user will be at some time in the future and making sure he/she will not hit anything there. These two behaviors are based on picking routines. The picking routine indicates whether there may be intersections with other objects. The collision avoidance behavior calculates where the user intends to go while navigating in the VE. If an object is in the way, it stops the user from going there (Figure 4.12(a)). The terrain following behavior checks whether the user is at the correct height above the ground (Figure 4.12(b)). The algorithms and computational aspects of the collision avoidance and terrain following behaviors are given in Appendix A.

(2) Space boundaries

This behavior checks whether user is inside the boundaries of the VE and does not let him/her go outside these boundaries. For example, it does not let the user go to locations outside the VE of Montreal.

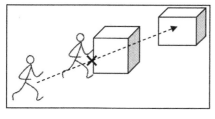 

(a) Collision avoidance　　　　　　(b) Terrain following

Figure 4.12 Collision avoidance and terrain following behaviors

4.5 USABILITY STUDIES OF NAVIGATION AND WAYFINDING

As discussed in Chapter 2, the usability criteria associated with interaction have been classified as: wayfinding (i.e., locating and orienting oneself in an environment); navigation (i.e., moving from one location to another in an environment); and object selection and manipulation (i.e., targeting objects within an environment to reposition, reorient and/or query). In this book, we focus on the usability testing of wayfinding and navigation in the UVEs. The test can be conducted based on the measurement of human performance for effective, efficient, and safe operation of the VR system.

4.5.1 Usability metrics

As Lampton et al. (2002) explained, when devising the performance measurement system for any VE systems, two levels of measures should be considered. The primary level will be determined by the specific application that the VE system is designed to address. The second level of measures will support the interpretation of the primary measures. The primary measures focus on outcome, indicating what the user accomplished in the VE system. The secondary measures help to interpret and elaborate on why performance was successful or not. Based on this approach, the time that it takes the users to orient themselves in the VE and find their targets can be measured as the performance measure of wayfinding. The performance testing of navigation methods in LBC-Infra is based on GISs because the VE in LBC-Infra is based on real locations. In order to define the measure for this test, the location of the users at every time step t_i while following a predefined path using different navigation methods are recorded as points in a GIS layer. It is assumed that the tolerance for following the path using the navigation control (mouse) to be a distance L from the path. Then the percentage of points outside a buffer surrounding the path and of width of L can be used as the measure of errors (Figure 4.13). In order to realize this type of testing based on user studies, a subsystem is designed in the GUI of LBC-Infra including the following functions: (1) Measure the time of the test; (2) Track the location of the user in the VE at every specific time step t_i; and (3) Record the previous locations in a new GIS layer for each user.

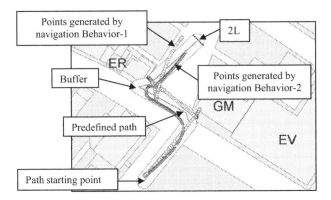

Figure 4.13 Sample of GIS layer containing user locations while following a predefined path

4.5.2 Result analysis of usability studies

The result analysis is an important part of performance measurement. The analysis of the results of the wayfinding test is done using the Analysis of Variance (ANOVA) method (Appendix B). For the analysis of the results of the navigation performance test, the users can be classified based on the percentage of errors. The number of users in each group can indicate how users perform using different navigation methods and which functionalities help users to perform better.

4.6 SUMMERY AND CONCLUSIONS

In this chapter, the interaction components of LBC-Infra are introduced. We focused on navigation and visualization component. The following conclusions about this chapter can be stated: (1) A taxonomy for navigation methods and support tools for engineering applications in UVEs is investigated; (2) Several computational issues and a well designed GUI for realizing the interaction components of LBC-Infra were investigated; and (3) A new GIS-based approach for usability testing of navigation in UVEs was proposed.

CHAPTER 5   IMPLEMENTATION AND CASE STUDIES

5.1 INTRODUCTION

Based on the discussion in Chapters 3 and 4, a prototype system is developed following

the framework architecture. In order to test the usability of some interaction components

of LBC-Infra, a new tool is developed for the usability testing of the navigation methods

using GIS. Three case studies of Concordia downtown campus and one major bridge in

Montreal (Jacques Cartier Bridge) are used to demonstrate the prototype system.

5.2 SELECTION OF DEVELOPMENT TOOLS

Java is a general purpose programming language with a number of features that make the

language well suited for use on the World Wide Web. Java is chosen to develop the

prototype system for four reasons: Object orientation, safety, simplicity, and breadth of

the standard library (Horstmann, 2004). Object orientation enables programmers to spend

more time on the design of their programs and less time coding and debugging.

Furthermore, graphics, user interface development, database access, multithreading, and

network programming are all parts of the standard library. The Java 3D Application

Programming Interface (Java 3D API) allows the programmer to describe the 3D scene

using coarser-grained graphical objects and defining objects for elements such as

appearances, transforms, materials, lights, etc. Compared with Open GL, the code is more

readable, maintainable, reusable, and easier to write (Selman, 2002).

Furthermore, MapObjects-Java Edition is available to build custom applications that

incorporate GIS and mapping capabilities (MapObjects-Java, 2006). MapObjects-Java

Edition helps the programmers build applications that perform a variety of geography-based display, query, and data retrieval.

The standard used for VR models is VRML. VRML is the most popular interoperability standard for describing interactive 3D objects and virtual worlds delivered across the Internet (Nadeau and Moreland, 1996).

The database of the 3D objects in the VE is designed with Microsoft Access to represent the information of all the objects such as buildings and bridges. Java Database Connectivity (JDBC) is used to access information stored in databases. The details about software requirements and installation guide of the prototype system are included in Appendix C.

5.3 BACKGROUND OF THE CASE STUDIES

Concordia downtown campus (Sir George Williams Campus) and Jacques Cartier Bridge in Montreal are chosen as the subjects of the case studies (Figure 5.1). Concordia is a large, urban university, in Montreal. The growth of Concordia's downtown campus has led to build two new buildings, the Integrated Engineering, Computer Science and Visual Arts Building and the new John Molson School of Business, as well as the recent acquisition of the Grey Nuns' Mother House.

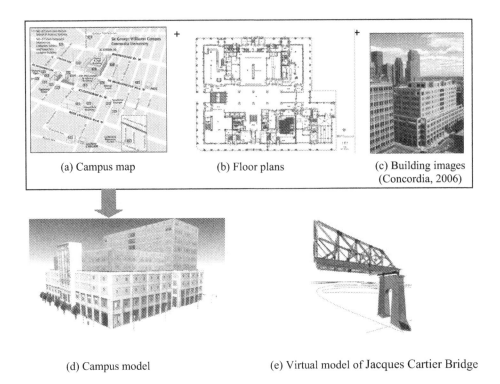

| (a) Campus map | (b) Floor plans | (c) Building images (Concordia, 2006) |

(d) Campus model (e) Virtual model of Jacques Cartier Bridge

Figure 5.1 The case study of an urban environment

5.4 IMPLEMENTING THE FRAMEWORK

5.4.1 Object-relational data model

The data model in the framework is an object-relational data model, which combines the relational data model with object-oriented development tools. Relational database management systems are still the norm. However, object-oriented modeling and programming tools are widely used in software engineering and can greatly enhance the quality of the software because of their flexible data structure. A good combination of the two approaches is the object-relational approach for database development, which can relate the information in the relational database with the data structure of building components as described in object-oriented programs (Object-Relational Mapping, 2004). For example, the data stored in the database about the structure of the VE are read automatically and a logical tree is created based on the structure. Figures 5.2(a) show an example of an object tree representing Concordia campus and its table representation, respectively. Through querying the database, the root of the tree is found and the root node is created. Then, queries are applied recursively to find other nodes based on the data stored in the table.

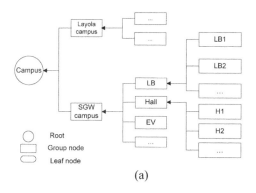

(a)

Group nodes	nodes
Root	Campus
Campus	SGW campus
Campus	Layola campus
SGW campus	Library Building
SGW campus	Hall Building
SGW campus	...
Library Building	LB1
Library Building	LB2
Library Building	LB3
...	...

(b)

Figure 5.2 Object tree: (a) Example of the object tree; and (b) its table representation

69

5.4.2 GIS integration

A GIS sub-system is created using MapObjects Java Edition (MapObjects-Java, 2006). The map includes several layers related to Montreal City, such as a boundary layer and other layers for the roads, blocks around the campus and Concordia buildings (Figure 5.3). In addition, the DEM model of the area was added to the scene graph. The DEM data source is the Canadian Digital Elevation Data (CDED). The CDED consists of an ordered array of ground elevations (recorded in meters) at regularly spaced intervals. To integrate this elevation information in our prototype system, the geographic coordinates have to be transferred to the world coordinates to match the 2D projected map of Montreal. Two CDED files, which cover the west and east parts of Montreal, are transferred to TIN model (more details are given in Appendix D).

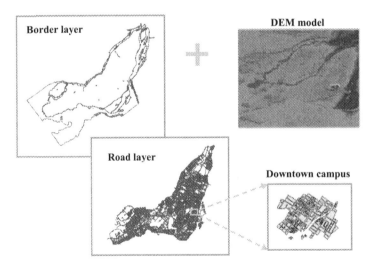

Figure 5.3 GIS information

5.4.3 Building the VR model of the campus

Following the modeling approach discussed in Section 3.2, the 3D virtual model of Concordia downtown campus is developed using the following data: (1) 2D CAD drawings of the buildings obtained from the Facilities Management Department of the university; (2) A digital map of the city of Montreal obtained from the municipality of Montreal; (3) A DEM of the city obtained from USGS website; (4) VRML library of small objects developed to embed in the 3D model, such as traffic lights, fire hydrants and street furniture; and (5) Orthogonal digital images of the facades of the buildings collected using a digital camera.

The 2D CAD drawings of the buildings are imported as a layer in ArcView (ArcView, 1996) and edited to create the outline of the buildings. The map of the area is imported in ArcView to create the block layer. The other layers including the image layer, object layer and tree layer are created using ArcView. The required attribute information of the layers is input in the attribute tables of the layers.

The images of the facades of the buildings are processed to cut the images so that they fit the building models and to make transparent parts of the images that are not necessary using Adob Photoshop software. Figures 5.4(a-b) shows two images of one side of a building that are edited and merged to create the image shown in Figure 5.4(c). The images and VRML files of street furniture are linked to the image and object layers, respectively. The GIS layers, images and 3D objects described above were integrated and translated into VRML. The translator application developed in Visual Basic uses a GIS

71

library to extrude the GIS shapefiles and create a number of VRML files that constitute

the virtual 3D model.

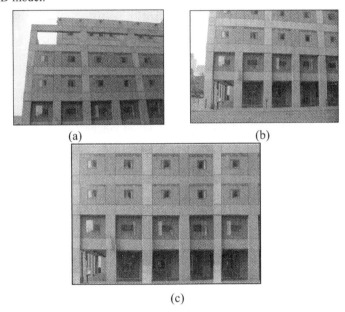

(a)   (b)

(c)

Figure 5.4 Example of images

5.4.4 Visualization of the VE

Virtual universes in Java 3D can be created from *scene graphs*. Scene graphs are

assembled from objects to define geometry, location, orientation, and appearance of

objects. Java 3D scene graphs are constructed from node objects using *BranchGroups* to

form a tree structure based on parent-child relationships (Figure 5.5). *TransformGroup*

objects can be constructed by applying *Transform3D* objects, which represent

transformations of 3D geometry such as translations and rotations (Walesh and

Gehringer, 2001). The following steps are used to visualize the VE:

(1) Loading VRML files: All the VRML files are imported into the Java 3D scene graph

   of the prototype system using VRML 97 API (J3d.org, 2006). The names of the buildings are read from the VRML file while loading and registered in a HashTable. Each building is assigned a unique ID equal to the name of the building in the database and registered in the HashTable, which has a key and an associated building ID. The HashTable allows quickly finding an existing building based on a key value. Because the VRML files are based on GIS data, it is located correctly in the 3D environment.

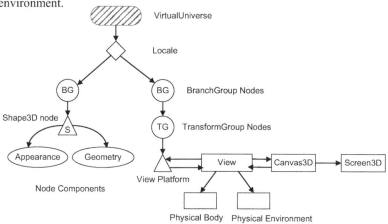

Figure 5.5 Scene graph (Walesh and Gehringer, 2001)

(2) *Loading CAD files*: A 3D CAD model of one floor of a building (the fifth floor of the EV building) is prepared and translated to VRML. In order to locate the model of the floor in the virtual 3D model, two points are identified in both models (i.e., two points on one edge of the building), and their coordinates are used to calculate the transformation matrix, including rotation, scaling and translation, from the local coordinate system of the CAD model to the global coordinate system of the virtual model. Another issue to be considered is the orientation of the coordinate axes used in

different visualization software. For example, the axis in the height direction is considered as the Y-axis in Java 3D and as the Z-axis in 3D Studio Max.

In addition, the 3D model of Jacques Cartier Bridge is added to the virtual model at the real location. The AutoCAD drawing of the bridge was acquired from the bridge management authority (Jacques Cartier and Champlain Bridge Incorporated). The 3D model of the bridge is created by converting the DWG file of the bridge into VRML.

(3) Loading floor plans: The DWG files can be loaded and visualized in a different panel using DWGLoader library (MapObjects Java, 2003). The 2D DXF files can be loaded and visualized in the virtual 3D model using DXFLoader library (j3d.org, 2006) which supports CAD files of various types (DXF, 3DS, OBJ, LWS). This loader scales the DXF by initial scale which fits the model into the view volume and also translates it by initial translation which centers the DXF model at the origin. In order to locate DXF model in the 3D model, the initial scale of the model is canceled and a point on one edge of the DXF model (P1) is translated to the origin (Figure 5.6 (a)). Then it is located in the virtual 3D model (Figure 5.6 (b)).

The geometry can be represented in three modes in the scene graph: solid, wire frame or point cloud (Figure 5.7 (a-b)).

(a) DXF model in the origin   (b) DXF model located in 3D

Figure 5.6 Locating 2D DXF model in 3D

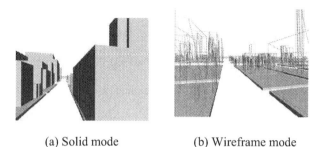

(a) Solid mode   (b) Wireframe mode

Figure 5.7 The different polygon modes

Road and building names can be displayed in the VE. The locations of the names are read

from the GIS layers. In order to find the location of the building names, the center point

of each building is read from the corresponding GIS layer. The road names are located in

the middle point between two intersections.

5.4.5 User interface development

The main user interface of the system is shown in Figure 5.8. On the left-hand side, a 3D

browser is displayed to show the VE. The user can navigate the VE and select an object

by picking that object using the mouse or other picking devices. Upon selection, the

object element is highlighted and the related information about the element is displayed.

Navigation buttons are positioned under the 3D browser to facilitate navigation especially

for novice users (more details are discussed in Section 4.4.2.1).

A logical tree of the campus structure is created automatically by reading the structure

data from the database. It is shown on the right-hand side of the user interface (Figure

5.8(a)). Each tree node has a check box, which highlights the selected building in the first level of the tree (buildings level). If the selected item is in the second level of the tree (floors levels) the 2D floor plan will be embedded in the 3D model.

Under the tree structure, there is a pane that allows user to increase and decrease the speed of navigation. The speed of navigation by default is 10km/s. The user can input a speed factor by which the current speed will be multiplied or divided and the new speed will be displayed.

The next pane is *Static viewpoints* pane that includes user defined and predefined viewpoints. Predefined viewpoints are listed which can navigate the user to some specific location in the VE. User defined viewpoints can be created during the navigation and given names by the user. In addition, the application provides predefined dynamic viewpoints (tours) which have a path through which the user can cycle to visit the world in a certain order.

The *Tracking pane* is designed to change the system mode to tracking mode. The *AR Mode button* will display the AR panel and starts video tracking. In this mode, using the *Full screen* button, the 2D user interface can be removed and 3D browser will be shown in full screen. The *Show 2D map* button will replace the tree structure of the campus by the interactive 2D map (Figure 5.8(b)). In this mode, the locations of the user in the VE will be tracked and will be shown on the 2D map as points. In addition, the location of the user can be changed in the 2D map and the viewpoint in the 3D environment can reflect this changes. The *Real time tracking* button will display the GPS tracking panel and starts GPS tracking.

The *System setting* panel is designed to change some of the default settings in the system (Figure 5.8(c)). In the *Rendering attribute* pane, several check boxes are listed that can be used to hide or show the visual cues that are discussed in Section 4.4, such as street names, DEM, etc. The *Tracking setting* contains controls to change some characteristics of the tracking system in different tracking modes. In the AR mode, the *automatic picking* can be disabled or enabled which allows to user to pick a building or an element in the VE just by looking at it continuously for a certain short time. In tracking the location of the user in the VR mode, the setting can be changed so that the current location of the user can be zoomed and centered in the 2D map pane. The *Navigation behavior* pane allows selecting a default behavior such as fly, orbit, drive and walk for moving around the scene (more details are given in Section 4.4.1.2). The *Polygon mode* pane allows changing the polygon mode for all objects in the scene graph. The *Field of view* pane includes a slider that can increase and decrease the field of view.

5.4.6 Usability study subsystem

In order to evaluate the wayfinding and the efficiency of navigation techniques, a simple subsystem is developed. This system includes a timer that measures the time while the test is in progress. As explained in Section 4.5, using this system, the locations of the user in the VE at a specific time step t_i are tracked and recorded in a GIS layer as points. In addition, the time and the viewpoint transformation information at each point are saved in the GIS layer. This information can be used to regenerate the path navigated by the user. A predefined path can be added to the VE. By pushing *Record locations* button, the system will ask the name of the GIS layer. It will create a new layer with this name at a specific location in the hard disk and will start recording the locations. The process will be stopped by pushing the *Stop record location* button (Figure 5.9).

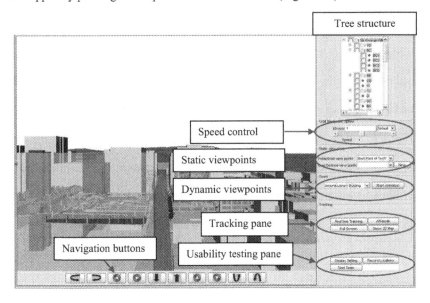

(a) Main window of the GUI

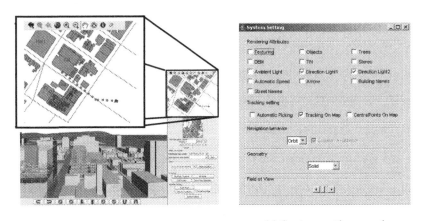

(b) 2D interactive map   (c) System setting panel

Figure 5.8 Screen shot of the user interface of the prototype system

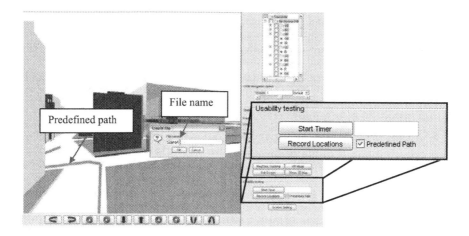

Figure 5.9 Usability testing subsystem

5.5 CASE STUDIES

The prototype system is used in three case studies: (1) Wayfinding and navigation in the VE; (2) Collection of inspection data for indoor applications; and (3) Collection of inspection data for outdoor applications.

In the first case study the navigation supports and methods provided in the prototype system are tested. In this case study the tracking is based on tracking the location of the user in the VE. The second and third case studies demonstrate applications of LBC-Infra in mobile engineering situations as discussed in Section 3.4. The second case study is based on tracking the location of the user inside a building using video tracking. The third case study is based on tracking the location of the user in the outdoors using GPS tracking. As discussed in Section 3.4, the location of the viewpoint in the VE is updated automatically based on the position received from the tracking devices and the navigation will be done automatically. But the navigation method and supports are helpful when user stops moving in the real world and looks around in VE to find objects of interest.

Therefore wayfinding and navigation methods are tested only in the first case study. The second and third case studies demonstrate how the proposed approach facilitates inspection tasks in mobile infrastructure management systems such as FMISs and BMSs.

5.5.1 Case study-1: Wayfinding and navigation in a VE

As reviewed in Chapter 2, the design of a VE must include appropriate visual cues and navigational aids, such as a map, to facilitate users' acquisition of spatial knowledge. Without the proper design of the navigational space and availability of tools to aid in exploring this space, the overall usability of a VE system will suffer resulting in ineffective and inefficient task performance. Wayfinding is possibly the most valid way to assess navigational knowledge. We conducted a usability test to evaluate the efficiency of the navigation techniques and wayfinding in the prototype system.

5.5.1.1 Procedure of the usability tests

The subjects were first given a detailed timeline of the test (Appendix E). The tests took approximately 25 minutes for each subject with the following steps:

(1) Training on how to use the system, and how to move in the VE;

(2) Exploration of the VE campus for few minutes;

(3) A detailed description of the conditions of the tests;

(4) Testing a wayfinding task under different conditions; and

(5) Testing the efficiency of navigation techniques by asking the user to follow a predefined path.

5.5.1.2 Wayfinding test

28 subjects participated in the test. The subjects were all volunteers and were recruited from engineering graduate and undergraduate students of Concordia University. The users were asked to navigate from one location within the campus to the main Hall building using four different types of navigation support: (1) with visual cues (landmarks, buildings' names and street names), (2) without any support, (3) with the interactive 2D digital map, and (4) with an animated arrow. As discussed in Section 4.7, the navigation time was considered as the metric in this test. Before starting the test, the users were given a few minutes to get familiar with the navigation methods and the environment. After that, they were asked to start the navigation from a specific location and to find the target under each condition. The time was measured while in progress. When the subjects recognized that they were in front of the correct building, the time was noted. They were then taken to the start point to start the navigation under another condition.

5.5.1.2.1 Results and Discussion

Three of the subjects could not complete the task and they were not considered in the analysis of the results. The average test times for the four types of navigation support were 22.36, 31.08, 20.07 and 22.76 seconds, respectively. The one-factor repeated measures ANOVA was performed on the test time data. In this analysis, the total variance has two components: between-subjects variance and within-subjects variance, because each subject gives more than one score. The effect of using the navigation supports was observed, $F_{(8, 59)}=2.82$, $p<0.01$ (Appendix F). This analysis proves that the new navigation support techniques significantly improve the efficiency of navigation in the VE. In addition, because the users were becoming more and more familiar with the environment as they were using one technique after the other, we could observe some improvement in their performance in latter tests.

5.5.1.3 Test of the efficiency of navigation methods

Eighteen subjects participated in this test. The subjects were all volunteers and engineering graduate students of Concordia University. In order to test the effect of collision avoidance on the efficiency of navigation in the VE, the subjects were asked to follow a predefined path drawn on the floor and not to deviate from it under two different conditions: (1) Using a navigation behavior with collision avoidance (Behavior-1), (2) Using a navigation behavior without collision avoidance (Behavior-2). The speed of navigation was adjusted to 10 m/s in the VE. The locations of the users at each time step of 100 ms were tracked and recorded in a GIS layer using the usability test subsystem (more details are given in Section 5.4.6).

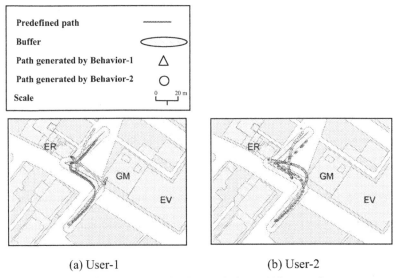

(a) User-1   (b) User-2

Figure 5.10 Navigation paths in usability test-2

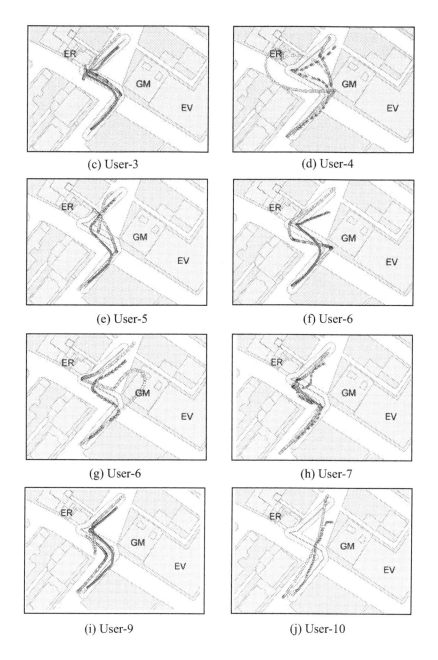

(c) User-3

(d) User-4

(e) User-5

(f) User-6

(g) User-6

(h) User-7

(i) User-9

(j) User-10

Figure 5.10 Navigation paths in usability test-2 (Continued)

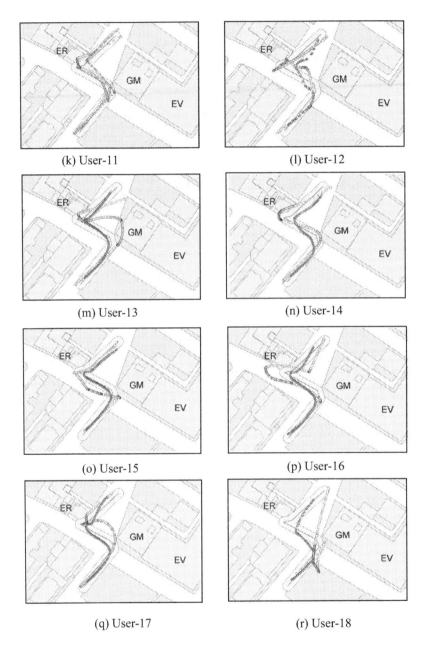

(k) User-11   (l) User-12

(m) User-13   (n) User-14

(o) User-15   (p) User-16

(q) User-17   (r) User-18

Figure 5.10 Navigation paths in usability test-2 (Continued)

Table 5.1 User performance in terms of percentage of errors and time of navigation

User #	Errors (%)		Navigation time (Sec.)	
	Behavior-1	Behavior-2	Behavior-1	behavior-2
1	0%	49%	17	31
2	12%	21%	18	28
3	0%	20%	11	27
4	13%	58%	16	28
5	6%	29%	22	30
6	15%	40%	19	34
7	2%	37%	25	38
8	32%	34%	20	37
9	0%	23%	15	29
10	21%	55%	25	32
11	19%	27%	20	33
12	15%	40%	22	35
13	0%	49%	19	28
14	28%	49%	23	27
15	12%	30%	20	31
16	4%	39%	24	36
17	8%	20%	18	25
18	11%	57%	21	29

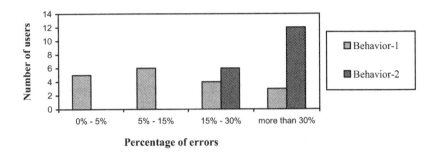

Figure 5.11 Performance of the users with and without collision detection

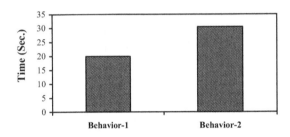

Figure 5.12 Mean time of navigation with and without collision detection

Figures 5.10 (a-r) show the generated paths by eighteen users during the test. The predefined path and navigation paths generated by different navigation behaviors are shown with different symbols.

5.5.1.3.1 Results and Discussion

As discussed in Section 4.5.2, the metric used to measure the performance in this test is the percentage of points outside a buffer with a certain width. The buffer is taken with 6m on each side of the path (L = 6 m). As shown in Figure 5.10, most of the users have orientation problems using navigation Behavior-2. For example User-1, User-2, User-3,

User-5, User-9, User-12 and User-17 repeated the same mistake when they wanted to turn right at the end of second segment of the path. They went through the wall and when they tried to come back to the road they moved in the backward direction.

Table 5.1 shows the percentage of the points outside the buffer for each user. Figure 5.11 shows a summery of users performance. The users are classified based on the percentage of errors and the number of users in each category is shown. The results clearly show that the users performed more efficiently using Behavior-1 (with collision avoidance).

Figure 5.12 shows the means of navigation time using Behavior-1 and Behavior-2 (20 sec. and 30.5 sec. respectively). This figure shows that collision avoidance has significantly decreased the time of navigation.

5.5.2 Case study-2: Indoor navigation

This case study, demonstrates the applications of LBC-Infra for indoor data collection for FMIS using video tracking (Mozaffari et al., 2005). As a basic example, the inspection routine was linked to the 3D model of the fifth floor of the EV building (Figure 5.13). The inspector can navigate in the floor; the viewpoint in the VE will be changed according to the current location of the user calculated by the video tracking system. As shown in Figure 5.13, the 2D map of the floor on the right hand side shows the current location of the user that can facilitate wayfinding in the floor. As explained in Section 4.4.1.2, the picking behavior can be used to mark the defects in the VE. When a defect is detected, the inspector can mark the defect in the VE and at a same time save the related information in the database. In addition, using the picking behavior, the inspector can get some information about different parts of the building such as column names and heights.

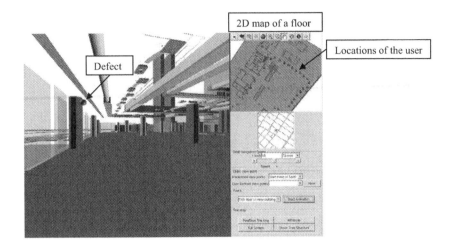

Figure 5.13 An example of inspection of a floor in FMIS

5.5.3 Case study-3: Outdoor navigation and tracking

This case study demonstrates the applications of LBC-Infra for outdoor data collection for BMS using GPS tracking (Hammad et al., 2004b). As an example, the inspection data collection was done based on the 3D model for the Jacques Cartier Bridge. Figure 5.13 shows the location-based visual inspection process of Jacques Cartier Bridge with the navigation and picking functions. Figure 5.14(a) shows the locations of the user collected during the inspection process using the GPS tracking. As shown in Figure 5.14(b), at the beginning of the inspection activities, virtual arrows automatically guide the inspector with a predefined inspection order according to the inspection plan. The arrows are created dynamically and inserted into the scene graph. The trajectory of the arrow is computed based on the present position of the inspector (obtained from GPS) and the location of the defect. Compared with the traditional manual data input, using the prototype system improved the efficiency of data collection with the help of navigation and directly marking defects on the bridge 3D model.

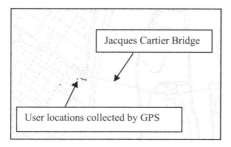

(a) The locations of the user collected by GPS

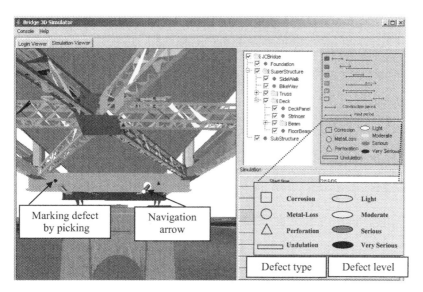

(b) Using 3D model of a bridge for inspection

Figure 5.14 An example of inspection of a bridge in BMS

5.6 SUMMARY AND CONCLUSIONS

This chapter described the implementation of the proposed approach and methods discussed in Chapters 3 and 4. The prototype system is developed by creating the models of an urban environment including Concordia downtown campus and the Jacques Cartier Bridge. The software development tools are selected to integrate several information technologies in the prototype system. The validity of the proposed framework is demonstrated by examining three case studies: Wayfinding and navigation in the VE, indoor navigation and outdoor navigation.

In the first case study, two usability tests were conducted to test the navigation supports and methods provided in the prototype system: wayfinding test and test of the efficiency of navigation methods.

28 subjects participated in the wayfinding test. They were asked to navigate from one location within the campus to the main Hall building using four different types of navigation supports: (1) with visual cues (landmarks, buildings' names and street names), (2) without any support, (3) with the interactive 2D digital map, and (4) with an animated arrow. The average test times for the four types of navigation support were 22.36, 31.08, 20.07 and 22.76 seconds, respectively. The results of this test showed that using navigation supports allow users to navigate more efficiently in the VE.

20 subjects participated in the test of the efficiency of navigation methods. In order to test the effect of collision avoidance on the efficiency of navigation in the VE, the subjects were asked to follow a predefined path drawn on the floor under two different navigation behaviors: with and without collision avoidance. The locations of the users at specific

time steps were recorded in a GIS layer. The metric used to measure the performance in this test was the percentage of points outside a buffer with a certain width. The results of this test clearly showed that the users have orientation problems without collision avoidance and collision avoidance has significantly decreased the time of navigation.

The second and third case studies demonstrated examples of applications of LBC-Infra in mobile engineering applications. In these two case studies, the location of the viewpoint in the VE is updated automatically based on the position received from the tracking devices and the navigation is done automatically. The second case study was based on tracking the location of the user inside a building using video tracking. The third case study was based on tracking the location of the user in the outdoors using GPS tracking. Compared with the traditional manual data input, using the prototype system improved the efficiency of navigation and data collection with the help of navigation and directly marking defects on the 3D models of infrastructures.

CHAPTER 6   NEW CHALENGES

As introduced in the literature review, UVE models have been useful for many urban applications such as planning and construction. However, in order to realize a 3D urban model, enormous amounts of time and money have to be consumed to design the model and to acquire the data for the model. Large-scale virtual models of urban environments can be created using commercial GIS software. However, these models focus only on the exterior shapes of buildings and do not include all the details necessary in engineering applications (e.g., design and scheduling information).

We introduced a data integration method for creating a system to generate VR models of an urban environment automatically by integrating GIS maps, CAD models, images of building facades, and databases of information related to buildings such as cost, scheduling and other data generated during the lifecycle of buildings. In addition, we proposed a framework for using these models to suit the requirements of mobile infrastructure management systems. This framework can be used by field workers to interact with geo-referenced infrastructure models and automatically retrieve the necessary information in real time based on their location and orientation, and the task context.

We introduced the interaction components of LBC-Infra and the different approaches for interaction with the 3D VE. In addition, we have presented a new taxonomy of navigation methods and support tools suitable for engineering applications in UVEs. Furthermore, a GIS-based approach for usability testing of navigation methods is proposed. This approach allows for the analysis of user behaviors during the navigation in VEs.

Three case studies of Concordia downtown campus and Jacques Cartier Bridge were used to demonstrate the prototype system and the applications using the new proposed approaches. In the first case study, usability testing of navigation is conducted based on user studies. In the remaining case studies, the prototype system is used to show the potential of using LBC-Infra by field workers in mobile situations.

Before we discuss the new challenges, below we have summarized the contributions of this work:

(1) A data integration method is developed for creating a system to generate VR models of an urban environment automatically by integrating GIS maps, CAD models, images of building facades, and databases of the data generated during the lifecycle of buildings.

(2) A framework was developed based on LBC for using these models to suit the requirements of mobile infrastructure management systems. Several issues necessary to realize the framework were discussed such as the GUI design and interaction methods.

(3) Based on the review of previous researches, a taxonomy of navigation methods and tools in virtual urban environments for engineering applications was defined.

(4) A GIS-based approach for usability testing of wayfinding and navigating methods in VEs was proposed. Several metrics were defined for measuring user performance while navigating the VEs.

(5) A prototype system, implemented in Java, was developed and several case studies in Montreal were used to demonstrate the feasibility of the above mentioned approaches and methods.

While pursuing this work, several challenges and limitations have been identified related to the requirements and the performance of the developed methods and techniques.

(1) The new automatic speed method needs testing and improvement.

(2) The proposed usability testing method can be used for regenerating the navigation path created by a user. This path can be used for further analysis of user behaviors during the navigation.

(3) The usability of the prototype system in mobile situation such as BMSs and FMSs needs more testing. Further development and testing in practical situations are necessary to improve the functionalities and usability of these systems.

(4) The usage of the proposed framework as a collaborative environment for field workers needs to be investigated.

(5) Creating 4D model of the UVE including the time dimension to show the changes of the urban environment during its lifecycle is another topic for future research.

REFERENCES

ASI website (2006). <http://www.anlt.com/>.

ArcView GIS (1996), The Geographic Information System for Everyone, Environmental System Reasearch Inistitute.

Ausburn, L.J. and Ausburn, F.B. (2004). Desktop Virtual Reality: A Powerful New Technology for Teaching and Research in Industrial Teacher Education, Journal of Industrial Teacher Education, Vol. 41, pp. 33-58.

Batty, M., Chapman, D., Evans S., Haklay M., Kuppers, S., Shiode, N., Smith, A. and Torrens P.M. (2001). Visualizing the City: Communicating Urban Design to Planners and Decision-makers, Planning Support Systems, CASA Working Paper 26, pp. 405-443.

Benyon, D. and Hook, K. (1997). Navigation in Information Spaces: Supporting the Individual, Human-Computer Interaction: Proceedings of Interact '97, pp. 39-46.

Bitmanagement Sotware website (2006). <http://www.bitmanagement.com>.

Bourdakis, V. and Day, A. (1997). The VRML Model of the City of Bath, Proceedings of the Sixth International EuropIA Conference, pp. 245-259.

Bowman, D. and Billinghurst, M. (2002). Special Issue on 3D Interaction in Virtual and Mixed Realities: Guest Editors' Introduction, Virtual Reality, Vol. 6, pp. 105-106.

Bowman, D., Kruijff, E., La Viola, J. and Poupyrev, I. (2000). The Art and Science of 3D Interaction, Tutorial notes from the IEEE International Virtual Reality 2000 Conference.

Bowman, D. (1999). Interaction Techniques for Common Tasks in Immersive Virtual Environments: Design, Evaluation, and Application, PHD thesis, Georgia University of Technology, < www.cs.vt.edu/~bowman/thesis/thesis_front.pdf>.

Burdea, G. and Coiffet, P. (2003). Virtual Reality Technology, Edition 2.0, John Wiley & Sons, Inc., Hoboken, New Jersey, ISBN: 0-471-36089-9.

CDED (2004). Canadian Digital Elevation Data Product Specification, Edition 2.0, Government of Canada, Natural Resources Canada, Center for Topographic Information, < http://www.cits.rncan.gc.ca/ >.

Charitos, D. and Rutherford, P. (1996). Guidelines for the Design of Virtual Environments, 3rd Virtual Reality Special Interest Group Conference, pp. 93-111.

Chen, J.L. and Stanney, K.M. (1999). A Theoretical Model of Wayfinding in Virtual Environments: Proposed Strategies for Navigational Aiding, Presence: Teleoperators and Virtual Environments, pp. 671-685.

CosmoPlayer website (2006). <http://ovrt.nist.gov/cosmo/>.

Cruz-Neira, C., Langley, R. and Bash, P.A. (1996). VIBE: a Virtual Biomolecular Environment for Interactive Molecular Modeling, Journal of Computers and Chemistry, Vol.20, pp. 469-477.

Darken, R.P. and Sibert, J.L. (1996a). Navigating in Large Virtual Worlds, The International Journal of Human-Computer Interaction 8 (1), pp. 49-72.

Darken, R.P. and Sibert, J.L. (1996b). Wayfinding Strategies and Behaviors in Large Virtual Environments, Human Factors in Computing Systems, CHI '96 Conference Proceedings, pp. 142-149.

Dlgv32 pro website (2006). <http://mcmcweb.er.usgs.gov/drc/dlgv32pro>.

Dodge, M., Doyle. S., Smith. A. and Fleetwood, S. (1998). Towards the Virtual City: VR & Internet GIS for Urban Planning, GIS Europe, Vol. 6, pp 26-29.

Durlach, N.I. and Mavor, A.S. (1995). Virtual Reality: Scientific and Technological challenges, National Academic Press, Washington, DC, ISBN: 0-309-05135-5.

ESC website (2006). <http://www.simcenter.org/ >.

ESRI website (2006). ArcView 3D Analyst, <http://www.esri.com/software/ arcview/extensions/3danalyst/ >.

Faust, N.L. (1995). The Virtual Reality of GIS, Environment and planning B: Planning and Design, Vol. 22, pp. 257-268.

Furnas, G.W. (1997). Effective View Navigation, Human Factors in Computing Systems: CHI '97 Conference Proceedings, Association for Computing Machinery, pp. 367-374.

Gabbard, J.L. and Hix, D. (1997). A Taxonomy of Usability Characteristics in Virtual Environments, <http://csgrad.cs.vt.edu/Bjgabbard/ve/taxonomy>.

Gabbard, J.L., Hix, D. and Swan, E. (1999). User-centered Design and Evaluation of Virtual Environments, IEEE Computer Graphics and Applications, pp. 51–59.

Gross, Z., and Kennelly, P.J. (2005). A Primer for Creating 3D Models in ArcScene, ESRI Journal, pp. 26-29.

Groneman, A.C. (2004). Toposcopy Combines 3D Modeling with Automatic Texture Mapping. Proceedings of the ISPRS Workshop on Vision Techniques Applied to the Rehabilitation of City Centers, pp 168-172.

Google Earth website (2006). < http://earth.google.com/ >.

Goldberg, S. (1994). Training Dismounted Soldiers in a Distributed Interactive Virtual Environment, U.S. Army Research Institute Newsletter, pp. 9-12.

Goza, S.M., Ambrose R.O., Diftler, M. A. and Spain, I. M. (2004). Telepresence control of the NASA/DARPA robonaut on a mobility platform, Proceedings of CHI 2004, pp. 623-629.

Halden Virtual Reality Center website (2006). <http://www2.hrp.no/vr>.

Hammad, A., Garrett, J.H. and Karimi, H. (2004a). "Location-Based computing for infrastructure field tasks", In Karimi, H. and Hammad, A. (editors), "Telegeoinformatics: Location-Based Computing and Services," CRC Press.

Hammad, A., Zhang, C., Hu,Y. and Mozaffari, E. (2004b). Mobile Model-Based bridge lifecycle management systems, Proc.,Conference on Construction Applications of Virtual Reality. ADETTI/ISCTE, pp. 109-120.

Herndon, K., van Dam, A. and Gleicher, M. (1994). The Challenges of 3D Interaction, SIGCHI Bulletin, pp. 36-43.

Hix, D. and Gabbard, J.L. (2002). Usability Engineering of Virtual Environments, Handbook of Virtual Environments: Design, Implementation, and Applications, pp. 681-699.

Hoffman, H. and Vu, D. (1997). Virtual Reality: Teaching Tool of the Twenty-first Century, Academic Medicine, pp.1076-1081.

Horstmann, C. (2004). Big Java, 2nd edition, John Wiley & Sons, Inc.

Isdale, J. (2003). Introduction to VR technology, IEEE Virtual Reality Conference (VR'03), p. 302.

Ishiguro, H., Capella, R. and Trivedi, M. (2003). Omnidirectional Image-based Modeling: Three Approaches to Approximated Plenoptic Representations, Machine Vision and Applications, Vol. 9, pp. 94-103.

Johnson, D. (1994). Virtual Environments in Army Aviation Training, Proceedings of the 8th Annual Training Technology Technical Group Meeting, Mountain View, pp. 47-63.

Kalawsky, R.S. (1999). VRUSE- Computerized Diagnostic Tool: for Usability Evaluation of Virtual Synthetic Environments Systems, Applied Ergonomics 30, pp.11-25.

Kaplan, E.D. (1996).Understanding GPS: principles and applications, Artech House.

Karimi, H. and Hammad, A. (editors) (2004). Telegeoinformatics: Location-Based Computing and Services, CRC Press.

Kato, H., Billinghurst, M. and Poupyrev, I. (2006). ARToolKit version 2.33: A Software Library for Augmented Reality Application, <http://www.hitl.washington.edu/artoolkit>.

Kaur, K., Maiden, N. and Sutcliffe, A. (1999). Interacting with Virtual Environments: an Evaluation of a Model of Interaction". Interacting with Computers, pp. 403-426.

Kessler, G.D. (2002). Virtual Environment Models, Handbook of Virtual Environments: Design, Implementation, and Applications, pp. 255-276.

Lampton, D.R., Bliss, J.P. and Morris, C. S. (2002). Human Performance Measurement in Virtual Environments, Handbook of Virtual Environments: Design, Implementation, and Applications, pp. 701-720.

Laurini, R. and Thompson, A.D. (1992). Fundamentals of spatial information systems, A.P.I.C. Series, Academic Press, New York, NY.

Levy, R.M. (1995). Visualisation of Urban Alternatives, Environment and Planning B: Planning and Design, Vol. 22, pp. 343-358.

McCauley-Bell, P. (2002). Ergonomics in Virtual Environments, Handbook of Virtual Environments: Design, Implementation, and Applications. Lawrence Erlbaum Associates, pp. 807-826.

MapObjects-Java web site (2006). <http://www.esri.com/software/mojava>

MapObjects Java edition (2003). Esri.

Milgram, P., and Kishino, F. (1994). A Taxonomy of Mixed Reality Visual Displays".
IEICE Trans. on Information and Systems, Vol. E77-D, pp. 1321-1329.

Martin, D. and Higgs, G. (1997). The Visualization of Socio-economic GIS Data Using
Virtual Reality Tools, Transactions in GIS, Vol. 1, pp 255-265.

Mozaffari, E., Hammad, A., and El-Ammari, K. (2005). Virtual Reality Model for
Location-Based Facilities Management Systems, 1st CSCE Specialty Conference on
Infrastructure Technologies, Management and Policy, Toronto.

Nadeau, D.R. and Moreland, J. (1996). The VRML Sourcebook, John Wiley & Sons,
Inc., publication, ISBN: 0471141593

Nielsen, J. (1993). "Usability Engineering." Academic Press, ISBN 0-12-518406-9

NUME project website (2006). < http://www.learningsites.com/VWinAI/CINECA/
VWAI_CINECA-BolognaVR.htm>.

Object-Relational mapping website (2006). <http://www.service-architecture.com/java-
databases/index.html>.

Panoram Technologies website (2006). <http://www.macs.hw.ac.uk/~hamish/
/9ig2/topic24.html>.

Pausch, R., Burnette, T., Brockway, D. and Weiblen, M. (1995). Navigation and
Locomotion in Virtual Worlds Via Flight into Hand-Held Miniatures". Proceedings of
SIGGRAPH in Computer Graphics, pp. 399-400.

Parallel Graphic website (2006). <http://www.parallelgraphics.com/products/cortona/>

Planet 9 Studio website (2006). <http://www.planet9.com/demos.html>

Potel, M. (1996). MVP: Model-View-Presenter, The Taligent Programming Model for C++ and Java, <ftp://www6.software.ibm.com/software/developer/library/mvp.pdf>.

Sadowski, W. and Stanney, K.M. (2002). Measuring and Managing Presence in Virtual Environments". Handbook of Virtual Environments: Design, Implementation, and Applications. Lawrence Erlbaum Associates, pp. 791-806.

Satyanarayanan, M. (2001). Pervasive computing: vision and challenges, IEEE Personal Communications, pp. 10-17.

Seffah, A. and Donyaee, M. (2005). Metrics and Measurement of Usability, 2nd edition of the International Encyclopedia of Ergonomics and Human Factors, Waldemar Karwowski (Editor). CRC Press/Taylor and Francis.

Selman, D. (2002). Java 3D Programming, Manning Publications Co.

Seren K., and Karstila K. (2001). MS Project – IFC Mapping Specification, Version 1.0, Eurostep [online]. Available from http://www.eurostep.fi/public/4D/ 4D_Mapping_V1_0.pdf [cited on October 18, 2005].

Shiode, N. (2001). 3D Urban Models: Recent Developments in the Digital Modeling of Urban Environments in Three-Dimensions, GeoJournal, Vol. 52, pp. 263-269.

Shneiderman, B. (1992). Designing the user interface, 2nd Edition. Addison-Wesley, Reading, MA.

Shiffer, M.J. (1995). "Interactive Multimedia Planning Support: Moving from Stand-alone Systems to the World Wide Web, Environment and Planning B: Planning and Design, Vol. 22, pp. 649-664.

Sinning-Meister, M. and Gruen, A. and Dan, H. (1996). 3D City Models for CAAD-Supported Analysis and Design of Urban Areas, Photogrammetry and Remote Sensing, Vol. 51, pp.196-208.

Smith, A., (1997). Realism in Modelling the Built Environment Using the World Wide Web, HABITAT, Vol. 4, pp.17-18.

Stanney, K.M. (2002). Handbook of Virtual Environment: Design, Implementation, and Applications. Lawrence Erlbaum Associates.

Stanney, K.M., Mollaghasemia, M., Reevesa, M., Breauxb, R. and Graeberc, D. A. (2003). Usability Engineering of Virtual Environments (VEs): identifying multiple criteria that drive effective VE system design, Human-Computer Studies, pp.447-481.

Strommen, E. (1994). Children's Use of Mouse-Based Interfaces to Control Virtual Travel. Proceedings of CHI Conference, pp. 405-410.

Sutcliffe, A. and Kaur, K. (2000). Evaluating the Usability of Virtual Reality User Interfaces". Behavior and Information Technology, IEEE Virtual Reality 2002 Conference, Vol. 19, pp. 415-426.

Tan, D.S., Robertson, G.G., and Czerwinski, M. (2000). Exploring 3D Navigation: Combining Speed-coupled Flying with Orbiting", CHI 2001 Conference on Human Factors in Computing Systems, pp. 418-425.

Tate, D.L., Sibert, L. and King, T. (1998). Virtual Environments for Shipboard Firefighting Training, Proceedings of the Virtual Reality Annual International Symposium, pp. 61-68.

Towell J.F. and Towell E.R. (1995). Internet Conferencing with Networked Virtual Environments, Internet Research: Electronic Networking Applications and Policy, Vol. 5, pp. 15-22.

Tromp, J.G., Steed, A. and Wilson, J.R. (2003). Systematic Usability Evaluation and design issues for collaborative virtual environments, Presence: Teleoperators and Virtual Environments, Vol. 12, pp. 241-267.

Turk, M. and Robertson, G. (2000). Perceptual User Interfaces, Communications of the ACM , pp.33-34.

Turner, J.R. and Thayer, J. (2001). Introduction to Analysis of Variance: Design, Analysis, & Interpretion, Sage Publications, Inc.

Umeki, N. and Doi M. (1997). Sensation of Movement in Virtual Space, Electronics and Communication, Vol. 80, pp. 74-82.

Urban Simulation Team at UCLA (2006). <http://www.ust.ucla.edu/ustweb/ust.html>

Virtual Reality Laboratory website (2006). Virtual Reality: a Short Introduction, Virtual Reality Laboratory, University of Michigan, College of Engineering. <http://www-VRL.umich.edu>.

Walesh, A. and Gehringer, D. (2001). Java 3D API Jump-start, Prentice Hall PTR.

Ware, C. and Slipp, L. (1991). Using Velocity Control to navigate 3D Graphical Environments: A comparison of three interfaces, Proceedings of the Human Factors Society 35th Annual Meeting, pp. 300-304.

Ware, C. and Jessome D. (1988). Using the Bat: a Six-Dimensional Mouse for Object Placement, IEEE Computer Graphics and Applications, pp. 65-70.

Webscap website (2006). <http://webscape.com/Worlds/tokyo.html>.

103

Zenrin website (2006). <http://www.zenrin.co.jp/products/Digital/L/Zi2HP/2/2.html>.

APPENDIX A: Collision avoidance and terrain following

The basic process of collision avoidance and terrain following is detecting where the user will be at some time in the future and making sure he/she will not hit anything there. These behaviors are based on a ray tracing method (j3d.org, 2006). The general idea behind ray tracing is to put the mathematical equation for the ray into the equation for the object and determine if there is a real solution. If there is a real solution then there is an intersection. Therefore the closest point of intersection can be found. Collision avoidance is concerned about what is around and terrain following is concerned what is below.

In collision avoidance, a finite line segment will be used. The start point is the current location of the viewpoint and the end point is the location of the viewpoint in the next frame. If the picking routine does not return anything, the viewpoint will not hit anything during the next frame. Therefore it will keep moving in the VE. Otherwise, the user will intersect with other objects. In this case, the location of the viewpoint will be adjusted to be right before the intersected object. The new location of the viewpoint is calculated as the sum of the vectors that describe the current position; the direction user is traveling and the adjustment for the next frame (Figure A.1).

In the terrain following, the expected location of the viewpoint in the next frame is calculated (P_1). Then, an infinite ray is projected from current location of the viewpoint (P_0) along the *down* vector to find the intersection point with the terrain (P_2). Based on the step-height (maximum changes in the terrain heights between P_0 and P_2) and the height-above-terrain (maximum changes in the viewpoint heights between P_0 and P_1), it will be decided what to do next:

- If the change in height from P_0 to P_1 is zero then the viewpoint in the next frame is adjust to be in the correct height above the terrain.

- If the step-height is less than the height-above-terrain, then the viewpoint in the next frame will be adjusted to be at the given distance above the terrain in point P_2. (Figure A.1 (a-b)).

- If the step-height is greater than the height-above-terrain, it will be considered as collision and the movement of viewpoint will be prevented.

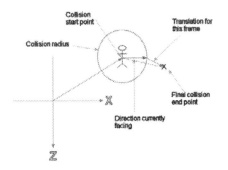

Figure A.1 Vectors contributing to find the end of the pick segment

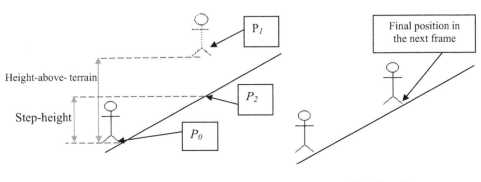

(a) Before adjusting the height   (b) After adjustment

Figure A.2 Terrain following

106

APPENDIX B: Analysis of variances (ANOVA)

ANOVA is a method used to compare the means of several groups of observations (Turner and Thayer, 2001). The analysis is based upon the assumption that the samples come from normally distributed populations with the same standard deviation. It is assumed that the variable of interest is normally distributed within each group and that each group has the same standard deviation for that variable. ANOVA is an extension of the difference of means test to more than two populations or samples. It allows researchers to examine multiple effects through sub-samples of a population. ANOVA does this by breaking down the total variance of any sample data set into component sources of variance. ANOVA involves the separation of the total variation found in three or more groups or samples into meaningful components: (1) Variability between groups, and (2) Variability within groups.

Between-group variability focuses on how the sample mean of each group differs from the overall or grand mean. Within-group variability measures the variation about the mean of each group and is used to estimate the variation within each group. The main goal in ANOVA is to see whether or not the variation between sample means is significantly greater than that within the samples themselves.

A greater variation between the samples than within the samples would suggest that the two groups came from different populations (i.e., a significant difference exists). The standard method of presentation of an ANOVA is through an ANOVA table (Table B.1)

Table B. 1 ANOVA Table

Source of Variation	Sum of Squares	Degrees of Freedom	Mean Square Variance	F test
Between Groups	SS_B	k-1	MS_B	
Within Groups	SS_W	N-k	MS_W	MS_B/MS_W
Total Variation	SS_T	N-1		

To examine an AVOVA table and make inferences regarding the multiple groups, a researcher uses a distribution of variance that is called the F test:

$$F = \frac{MS_B}{MS_W}, \text{ Where}$$

MS_B: Difference between the group means and the grand mean

MS_W: Variation within groups

The shape of an F distribution depends on both the denominator and numerator degrees of freedom and a given α level. The following conclusions can be drawn from the calculated F values: (1) If the three or more variances were similar then the F value would be near one and lies within the body of the curve, and (2) If the between-groups variance is much greater than the within groups variance, the calculated F value falls on the skewed tail of the distribution.

APPENDIX C: Software requirements and installation guide of the prototype system

Software requirements:

(1) Borland JBuilder 2005 Enterprise: used to develop the prototype system of BMS;
(2) MS Access (MS Access XP): used to store the lifecycle data of the bridge;
(3) ArcGIS (ESRI 2004): used to develop GIS application;
(4) Netica: used to develop the prototype of learning-based BN model;
(5) Windows XP: used as the operation system.

Installation guide:

1. Copy four folders to corresponding driver and change the associated code in the project to match the driver path. The contents in these folders include:
   - *Currproject* or *infra_project* (The folder includes all codes of our project)
   - *Javasoft* (The folder includes all libraries which are required in our project)
   - *Bridgere* (The folder includes all 2d information)
   - *BridgeResources* (The folder includes all 3d information and models)

2. Click Start->Control Panel-> Administrative Tools->Database Source.
   And add ODBC data source as below:
   - Microsoft Access Driver: Name: bridge, Location: C:\ BridgeResources\db1.mdb

   (No password for data source is required, so just leave the password as blank.)

3. Launch Jbuilder, open the project.jpx file. Then click the menu of Jbuilder: Project->Project Properties. Click the tab "Required Libraries". Then edit or add the path of libraries as below:

VRML97 (VRML File Loader API)
   - Download the library from: https://j3d-vrml97.dev.java.net/ and install it. Or you can get it from the path: D:/javasoft/loaders/vrml97.jar
   - Edit or add the vrml97.jar file to the path of VRML97 library.

Jess61p4 (Java Expert System Shell API) - Version 6.1
   - Download the library from: http://herzberg.ca.sandia.gov/jess/ and install it. Or you can get it from the path: D:/javasoft/Jess61p4/jess.jar.
   - Edit or add the jess.jar file to the path of Jess61p4 library.

Jdk3D (Java development Kit 3D) - Version 1.3.1
   - Download the library from: http://java.sun.com/products/java-media/3D/downloads/index.html and install it. Or you can get it from the path: D:/javasoft/ JRE/1.3.1_09/lib/ext
   - Edit or add all jar files under the Jdk3d directory to the path of Jdk3D library.

109

Javacomm (Java Communications API) - Version 3.0
- Download the library from:
  http://www.sun.com/download/products.xml?id=43208d3d and install it. Or you can get it from the path: D:/Javasoft/commapi/
- Edit or add the comm.jar file under the commapi directory to the path of Javacomm library .

DXFLoader (DXF file loader API) - Version 1.0
- Download the library from: http://www.johannes-raida.de/index.htm?cadviewer and install it. Or you can get it from the path: D:/Javasoft/DxFloader/
- Edit or add the dxfloader.jar file under the dxfloader directory to the path of DXFLoader library.

MOJ (MapObject Java API) - Version 2.1
- Download the library from: http://www.esri.com/software/mojava/ and install it.
- Edit or add all jar files under the directory C:/ESRI/MOJ21/lib to the path of MOJ library. Please also add the tutsource.jar and tutorial.jar files that are available at the directory MOJ21\Samples\Tutorial.

Netica (Netica Java API) - Version 2.17
- Download the library from: http://www.norsys.com/netica-j.html#download and install it. Or you can get it from the path: D:/Javasoft/ NeticaJ_Win/bin
- Edit or add the path of the directory NeticaJ_Win/bin to the path of Netica library.

JMF (Java Media Framework API) - Version 2.1.1
- Download the library from: http://java.sun.com/products/java-media/jmf/2.1.1/download.html
  and install it. Or you can get it from the path: D:/Javasoft/ jmf211e_scst/build/win32/lib
- Edit or add the path of the directory jmf211e_scst/build/win32/lib to the path of JMF library.

JARToolkit (Java ARToolkit API) - Version 2.0
- Download the library from: http://jerry.c-lab.de/jartoolkit/ and install it. Or you can get it from the path: D:\JavaSoft\JARToolkit Dlls
- JARToolkit need the dll files as below:
  - JARFrameGrabber.dll
  - JARToolkit.dll
  - JARVideo.dll
  - libARvideo.dll
  - libARvideod.dll
  - msvcr70.dll

Make sure you put these dll files in your JARToolkit Dlls directory, then click start-> Control Panel->System->Advanced->Environment Variables, please add the path of JARToolkit Dlls directory to the option"path" of User Variables.

Notes:

* When you use a different account in the same computer, you have to separately set the path of libraries for every account. It means you cannot just set the libraries for all accounts at the same time.

* If your code cannot be compiled following the above instructions, please carefully check the installation instruction. If the error information is about Java 3D, that means your computer does not have Java 3D. You can get the Java 3D package from the Java 3D folder.

Table C. 1 Summary of libraries used in the prototype system

Library	Description	Source	Version
VRML97	VRML File Loader API	https://j3d-vrml97.dev.java.net/ D:/javasoft/loaders/vrml97.jar	
Jess61p4	Java Expert System Shell API	http://herzberg.ca.sandia.gov/jess/ D:/javasoft/Jess61p4/jess.jar	6.1
JDK 3D	Java 3D API	http://java.sun.com/products/javamedia/3D/downloads/index.html D:/javasoft/ JRE/1.3.1_09/lib/ext	1.3.1
Javacomm	Java Communication API	http://www.sun.com/download/products.xml?id=43208d3d D:/Javasoft/commapi/	3.0
DXFLoader	DXF File Loader API	http://www.johannes-raida.de/index.htm?cadviewer D:/Javasoft/DxFloader/	1.0
MOJ	MapObject Java API	http://www.esri.com/software/mojava/ C:/ESRI/MOJ21/lib	2.1
Netica	Netica Java API	http://www.norsys.com/netica-j.html#download D:/Javasoft/ NeticaJ_Win/bin	2.17
JMF	Java Media Framework API	http://java.sun.com/products/java-media/jmf/2.1.1/download.html D:/Javasoft/jmf211e_scst/build/win32/lib	2.1.1
JARToolKit	Java ARToolKit API	http://jerry.c-lab.de/jartoolkit/ D:\JavaSoft\JARToolkit Dlls	2.0

APPENDIX D: Digital Elevation Model of Montreal

The DEM data source is the Canadian Digital Elevation Data (CDED). The CDED consists of an ordered array of ground elevations (recorded in meters) at regularly spaced intervals (CDED, 2004). The grid spacing is based on geographic coordinates at a maximum and minimum resolution of 0.75, 1.5 and 3 arc seconds for the scale of 1:50,000, and 3, 6 and 12 arc seconds for the scale of 1:250,000 respectively, depending on the latitude. In our prototype system, the 1: 50,000 CDED is used. For this scale, the grid spacing is always 0.75 arc seconds along a profile in the South-North direction and varies from 0.75 to 3 arc seconds in the West-East direction, depending upon the three geographic areas (Figure D.1). Montreal is in the A area, where the grid spacing in the West-East direction is 0.75 arc seconds. To integrate this elevation information in our prototype system, the geographic coordinates have to be transferred to the world coordinates and match the 2D map of Montreal. Two CDED files, which cover the West and East parts of Montreal, are transferred to TIN model, and then projected to Modified Transverse Mercator (MTM) file using the parameters from Table D.1.

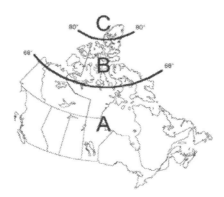

Figure D.1 Coverage of the three geographic areas

112

Table D. 1 50,000 CDED cell coverage according to the three geographic areas

GEOGRAPHIC AREA	LATITUDE		SPACING (latitude and longitude in arc seconds)		SPACING (in metres, approximate)		CELL COVERAGE (latitude - longitude)		
	from	to	lat.	long.	N.-S.	E.-W.			
A	—	68°	0.75" x	0.75"	23 m x 16-11 m		15'	x	15'
B	68°	80°	0.75" x	1.5"	23 m x 17-8 m		15'	x	30'
C	80°	90°	0.75" x	3"	23 m x 17-8 m		15'	x	1°

In addition, a color coding is defined to display different areas of the terrain according to the elevation values. Most of the elevation values of Montreal are below 100 and different ranges are defined to approximately represent the terrain following the color coding given in Dlgv32 pro software (2006).

APPENDIX E: Instructions and experimental timeline of usability test

Welcome to the Infrastructure Management Lab (Infra Lab). For the next 25 minutes you will be participating in a usability study examining what is the most effective manner to expose someone to a new virtual environment. For this study you will explore the virtual model of Concordia Downtown Campus. The following is a timeline of what you will be asked to follow.

(1) You will be taken to the intersection of Guy Street and St. Catherine (besides the EV building). For the first 10 minutes you will be asked to find the entrance of the Hall Building in the campus under the following conditions:

> (a) Using visual cues (DEM, Textures, Building names and Street names);
>
> (b) Without any navigation assistance;
>
> (c) Using animated arrows; and
>
> (d) Using the 2D map.

You should use the Left mouse button to navigate.

(2) You will be taken again to the intersection of Guy Street and St. Catherine (besides the EV building). For the next 10 minutes you will be asked to follow the path laid out for you (green path). It is very important that you follow this path. You will do this task under two different conditions (each about five minutes):

> (1) Navigation with collision avoidance ; and
>
> (2) Navigation without collision avoidance.

(3) If you have any questions, please ask them before you start the test or after you finish. You are free to make any comments you want during the exploration period.

Thank you for your participation.

APPENDIX F: Single factor ANOVA repeated measurement

Table F.1 Navigation time under four different conditions

User#	Navigation time (Sec.)			
	With cues	With no assistance	With map tracking	With arrow
1	16	20	18	15
2	23	28	17	19
3	29	42	26	27
4	24	43	19	28
5	16	19	15	17
6	16	26	14	20
7	23	31	19	22
8	17	23	16	18
9	21	28	17	20
10	17	24	17	20
11	19	25	16	18
12	28	38	30	35
13	30	38	29	28
14	19	24	18	21
15	32	38	26	28
16	42	48	40	42
17	20	28	17	25
18	21	26	22	26
19	22	31	18	20
20	24	39	21	22
21	19	31	17	18
22	15	30	16	19
23	26	39	21	23
24	15	23	14	16
25	25	35	19	22

Table F.2 ANOVA analysis result

Source of Variation	SS	df	MS	F	P
Between-Subjects	18743.99	23	-----		
Within-Subjects					
Conditions	5584.194	2	2792.097	17.04953	< 0.01
Subjects X Conditions SS	7533.139	46	163.7639		

APPENDIX G: Examples of navigation tools in commercial VRML browsers

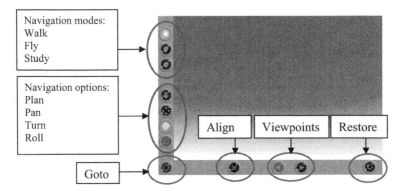

Figure G.1 Cortona VRML client (Parallel Graphic, 2006)

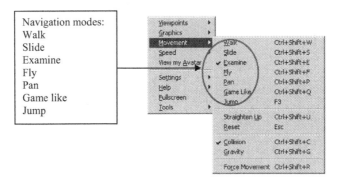

Figure G.2 Contact VRML (Bitmanagement Sotware, 2006)

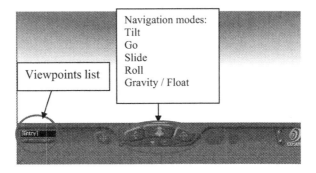

Figure G.3 CosmoPlayer (CosmoPlayer, 2006)

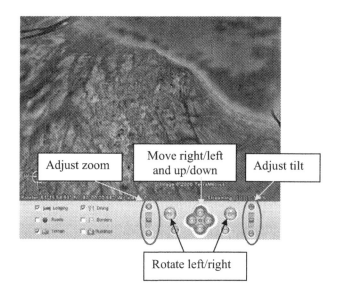

Figure G.4 Google Earth (Google Earth, 2006)

Lightning Source UK Ltd.
Milton Keynes UK
UKOW06f2157050914

238166UK00004B/132/A